Diary of an Apprentice Astronaut

Diary of an Apprentice Astronaut

SAMANTHA CRISTOFORETTI

translated from the Italian by Jill Foulston

ALLEN LANE
an imprint of
PENGUIN BOOKS

ALLEN LANE

UK | USA | Canada | Ireland | Australia
India | New Zealand | South Africa

Allen Lane is part of the Penguin Random House group of companies
whose addresses can be found at global.penguinrandomhouse.com.

First published in Italy by La Nave di Teseo 2018
First published in Great Britain by Allen Lane 2020
001

Copyright © Samantha Cristoforetti/ESA, 2020
Translation copyright © Jill Foulston, 2020
Illustrations by Jessica Lagatta in consultation with Samantha Cristoforetti

The moral rights of the author and translator have been asserted
Author's royalties donated to UNICEF

Set in 12/14.75 pt Dante MT Std
Typeset by Jouve (UK), Milton Keynes
Printed and bound in Great Britain by Clays Ltd, Elcograf S.p.A.

A CIP catalogue record for this book is available from the British Library

ISBN: 978-0-241-37138-1

To Kelsi Amel, who came as a stargazer

Contents

Dear reader,

This is the story of a journey. It was my journey, but I do not own this story. I'm entrusting it to you. Take care of it, though don't worry if it blends with your imagination, if it's nourished by your feelings, if it comes out transformed. That's how it should be.

Everything you're about to read is true, insofar as I have managed to avoid making mistakes. I'm afraid there may be a few, so I apologize to you here and now. I can assure you that I relied on my memory as little as possible, and only when I felt I really could.

Everything you're about to read here actually happened – but you won't find everything that happened in these pages. Out of discretion, because I only talk about the good I've found in others, and because I don't see how my private life can be of any interest. All the same, I haven't left anything out in an attempt to please you. I'm not trying to entertain you or teach you anything. I'm telling you this story – the whole of it – in the same way I would be telling it to a friend who was going with me on a long train journey. You're free to decide what to make of it.

I'll take you along during all my preparations, in the classrooms and simulators, in the swimming pool and the centrifuge, during emergency and survival drills – everywhere I go, suitcases in hand, all over the world. This is not a textbook, but we'll be picking up the many elements of an astronaut's trade along the way, a bit like collecting pebbles casually on a walk in the woods. When we arrive at the launchpad, I promise you, we'll be ready to fly into space.

This is a journey we'll make together. You and I, apprentice astronauts.

Samantha

To know is not to pull apart nor to explain. It is to attain vision. But in order to see, we must first participate, and it is a hard apprenticeship.

Antoine de Saint-Exupéry, *Flight to Arras*

Lightness for me is related to precision and definition, not to the hazy and haphazard. Paul Vale'ry said, "One must be light like the bird, not like the feather."

Italo Calvino, *Six Memos for the Next Millennium*

I.

We are a ball of fire in dizzying descent towards the planet, an incandescent wound in the thin atmosphere that envelops the Earth. We're slicing through the air at stupendous speed, air that gets so hot it becomes plasma. We are a falling star: if it were night, someone, somewhere would be wishing on us.

The little window on my left begins to darken on the outside, while shades of fiery orange dance around us. We're shrouded in ionized molecules of air, and they are blocking radio communication: just before we rejoin the Earthlings, our virtual umbilical cord, which has kept us connected to Mission Control in Moscow, is severed. It's almost a new birth: in a few minutes we'll emerge from the plasma, the parachute will open, and we'll make contact with the search and rescue helicopters of the Russian military already airborne over the Kazakh steppe. If everything goes to plan, we'll be reborn as Earthlings.

For 200 days, we've been extraterrestrial humans, orbiting at an altitude of 400 kilometres aboard the International Space Station. We've flown over every part of the Earth except the extreme north and south of the planet, every sea, every mountain and desert, every city and every port. We've witnessed a perpetually renewing, silent spectacle, all of it played out for billions of years before we humans laid eyes on it. As residents and guardians of humanity's outpost in space, we've inhabited weightless bodies and held weightless objects. We've felt the liberating and invigorating power of dreams come true, along with the responsibility of deserving, every day, a privilege reserved for few: representing humanity in space, the final frontier.

Only three hours ago we released the hooks that anchored us to the Space Station. Only half an hour ago we switched on the engine of our small Soyuz spaceship, slowing its unrelenting race around the planet. Just a bit, but enough to deliver us to the embrace of the atmosphere, which is now, little by little, relieving us of our extraordinary speed. The effect of this deceleration presses us into our seats: we're about to reach a first peak of 3.6 Gs, or 3.6 times our normal weight. Nothing out of the ordinary when we experienced it in the centrifuge on Earth, but fierce pressure after 200 days of absolute lightness.

My attention is split between my breathing, which is becoming increasingly difficult, and the display in front of me, which shows the parameters of our re-entry into the atmosphere. Everything is working perfectly: the on-board computer is automatically configuring a trajectory towards a location in the sky 10 kilometres over Kazakhstan, where the parachute will open. The voice of our commander comes through my earphones, calm and clear despite the pressure on his throat. He is making a report on the descent for the record: we're close to the nominal trajectory, and our deceleration has reached a second peak of 4.1 Gs and is now falling. No one can hear us right now, cocooned as we are in plasma, but already, many eyes are trained on the moment of our arrival.

The Astrey crew is coming home. We are the ones falling from the sky.

2.

'I wonder,' he said, 'whether the stars are set alight in heaven
so that one day each one of us may find his own again.'
 Antoine de Saint-Exupéry, *The Little Prince*

Istrana Air Base, Treviso, 18 May 2009

It was the most important phone call of my life – and I missed
it because I took a few extra minutes in the shower. It's already
evening, and I've lost hope of being contacted by ESA, the European Space Agency, today. But it couldn't have been anyone else:
an unknown French number, no message on the answerphone.
In my Spartan little room I sit down on the squeaky bed, with
its blue Air Force bedspread, and I take a few deep breaths to stop
the pounding of my heart, which seems about to burst from my
chest. What should I do? Wait for them to call back? Call back
myself? I've waited for this call for weeks, getting more and more
wound up by the day. In Italy, France, Germany, the United Kingdom, Denmark, and Finland, nine other young Europeans are
waiting, just like me, all of them gripped by the same anxiety.
We are the final ten applicants for the position of astronaut in a
selection process that has lasted more than a year. We expected
it to end a few months ago, after twenty-two of us attended an
interview with a group of ESA's senior managers. But we had a
final surprise last month. Instead of the expected yes-or-no outcome, an end to our agonizing uncertainty, there was another call
from France, with an invitation to one more test: a final interview
with the director general. Only ten of us were involved. Thousands of applicants have already been eliminated in previous selection phases, but a rejection at this stage would be devastating,

just when the dream has taken on substance and is no longer a wisp of will and imagination.

In the past few days the tension has become almost unbearable. The press conference and introduction of the selected applicants is scheduled to take place in Paris the day after tomorrow. Who would have imagined that we'd still be waiting for the final outcome at nine in the evening, two days before? Fortunately, we belong to a connected generation: either directly or indirectly, we are all in touch and we reassure each other that no one else has had any news.

I'm still sitting on my bed, not sure what to do, when an email drops into my inbox. I jump. Subject: ESA Astronaut Selection. I open it immediately, expecting only a number and a time to call back. Instead, one simple sentence, dry and technical, dissolves my tension. Every fibre in my body, my entire being relaxes. I don't cheer, I don't laugh, I don't cry. I'm pervaded by a feeling of quiet joy, and I relish a profound relief, in silence. There's no past, no future, only the dazzling present. The entire universe has stopped to give me a benevolent smile.

I reply to the email and then confirm by phone: I will, of course, be more than happy to come to Paris the day after tomorrow. And I understand I should say nothing so as not to compromise the press conference. I expected such a request, and made my excuses to friends a few days ago, letting them know that they would probably learn the outcome of the selection via the media. I call only my parents, who've watched me dream of becoming an astronaut ever since childhood, created opportunities for me, helped to make that dream a reality. I hear my own relief in their voices.

After that, I write to the remaining applicants with whom I'm in contact. If I'm meant to stick to the letter of the instructions about confidentiality, I probably shouldn't. But we promised to keep each other in the loop, and the shared torture of waiting over the past months has cemented our bond to the point where it seems unthinkable not to keep my promise. Late that night, two of them write to tell me that they haven't yet received any communication. No one says so, but it's clear that they probably have not been chosen. Sometimes only a thin and arbitrary line separates heady success from crushing disappointment.

Before going to bed, I look up at the patch of sky I can see through my window. One day, when I'm on board the International Space Station, I, too, will be just a bright dot up there. I can hardly believe that I've actually got to this place in my life. With a mixture of talent, hard work and extraordinary good luck, I've been able to realize something that's almost impossible. Life gave me a powerful yet insidious dream, because becoming an astronaut is terribly unlikely. Now, however, the path towards space is open. I may have to wait many years but, sooner or later, I trust that a rocket will be waiting for me on the launchpad.

I go to sleep thinking: today I found my star.

———

It had all begun more than a year earlier, in the changing room at a swimming pool close to the Istrana air base, northwest of Venice, where I was stationed. After my evening swim, I was surprised by an unexpected flood of emails, texts and calls, all with the same exciting news: according to various sources, the European Space Agency was about to start the selection process for a new group of astronauts. Although no one said it outright, there was one clear message flowing through the words of my friends and family: This is it! Here was my chance to realize my life's dream.

I reacted cautiously, tamping down my excitement to avoid the disappointment of a false alarm. What were the requirements? Was I sufficiently qualified? I worried that the moment had come too soon. At thirty-one years old, I was younger than most European or American astronauts when they were selected, and I had barely begun my career as a military pilot. I was a long way from qualifying as a test pilot, something I hoped to do one day, since it's always been a privileged gateway to space. I still needed a few years.

On the other hand, by the time the next chance rolled around I'd probably be above the acceptable age. The selections in the United

States and Russia take place every few years, but in Europe they are very infrequent, a once-in-a-lifetime opportunity for anyone aspiring to be an astronaut. The last one had been ten years before, in 1998, when I was still attending the Technical University of Munich. It was the year they launched the FGB, the first module of the International Space Station, and I was finishing up my second-year exams in mechanical engineering and looking forward to a specialization in aerospace for the next three years. I was twenty-one years old, a sadly significant age, since it was the cut-off for entering the Italian Air Force Academy as an aspiring pilot. I'd long contemplated this career and eagerly waited for the Italian armed forces to open up to women, so far in vain. There have been many favourable circumstances in my life, but perhaps the most outrageously lucky was this: the very next year, in a turn of events both surprising and unexpected, the law introduced voluntary military service for women, temporarily raising the age limit for female applicants and opening up an opportunity that seemed to have disappeared. So, despite the practical difficulties of travelling to Italy to take part in the admission tests and screenings, it was an obvious choice for me to apply in my last year of university. I was in Moscow then, grappling with my thesis on solid rocket propellants. At first I was doubtful about my chances for success, but one day in August I received a telegram summoning me to the Air Force Academy. I couldn't believe it: I'd come first in the admissions competition.

That's how I ended up in the changing room at the swimming pool all these years later, thinking back to the somewhat risky decision that saw me starting a new bachelor's degree course all over again. Me, a twenty-four-year-old, in the same class with students just out of high school and much younger than me. Spurred on as I was by the announcement of the imminent selection of astronauts, I questioned the wisdom of that choice, which had cost me dear in terms of freedom, relationships and alternative futures. Of course, it would bear fruit, but only slowly. After the Air Force Academy, I had earned my fighter pilot wings in the US and then I had taken the generic lead-in fighter training in Italy. Only now, after long months of waiting, a spot had opened up for me at the Operational Conversion

Unit in Amendola, near Foggia. There, I'd begin the course specific to the aeroplane assigned to me, the AM-X fighter bomber.

I was happy, sure, but I'd far rather have faced the astronaut selection with my course under my belt! So many unknowns sprang to mind as I returned from the swimming pool to my little room at the air base. Maybe ESA would immediately reject me as an applicant, judging that I didn't have enough flying hours. Or the Air Force wouldn't authorize me to take part in the selection process since I was still in training. And what an ironic twist of fate in the unlikely coincidence that, just as I was about to embark on a very difficult course that would demand my utmost dedication and focus, ESA should announce a new astronaut selection: I would need to give this chance of realizing my life's dream everything I had. How would I balance the two objectives together for an entire year? How would I give my all to one of these goals without compromising the other?

A few weeks later, I transferred to the Amendola air base, full of misgivings, to settle into yet another furnished room. From Bolzano to Minnesota, Munich to Toulouse, Moscow to Naples, Texas to Treviso, I had been travelling light since my teens, with just a few bags, which allowed me to erase all traces of my presence with ease, and temporarily put down roots in another place. I've never suffered from nostalgia, though I would many years later, when I returned to Earth from space.

Still, I stubbornly kept in touch with many friends I had met in those years of nomadic life. When I shared with them the news and the hopes of the moment, I received in return many heartfelt messages of encouragement, which made it sound like becoming an astronaut was now just a mere formality. I could only smile at their words and take it all with a pinch of salt, knowing full well that in friendship objectivity can be obscured by affection.

On the other hand, I took great comfort in a conversation I had with Maurizio Cheli, a former test pilot with the Italian Air Force and veteran of a Space Shuttle mission. We were put in contact through a shared acquaintance, and during a long conversation, which I'll always remember with gratitude, he listened to me carefully and offered precious advice and encouragement. He even declared at one point, 'You're the perfect candidate!' But he quickly added a sobering remark: 'Only a handful of people will be selected. This will be a dogfight!!' Indeed so . . . Who knew how many 'perfect candidates' from the then seventeen member states of ESA were queueing up to participate in the selection?

I'd been in Amendola for a week, buckling down to my theoretical course on the AM-X systems along with three wonderful classmates, when on 19 May the website astrosel.esa.int went live. To get a password allowing you to access the site, you had to send a medical certificate confirming your fitness to fly as a private pilot: a clever way of eliminating applicants who were clearly unfit.

Once you had the password, you could access a long questionnaire asking detailed information about your school and university education, as well as your professional background. There was also space to include flying hours or scientific publications, language skills, level of manual dexterity, experience in public speaking, volunteering, hobbies or sports – with a particular emphasis on activities such as diving, caving or parachuting. Finally, I had to provide a brief explanation of my motivation and my idea of an astronaut's duties. I had to describe my strengths and weaknesses and include any additional information that supported my application.

I spent many hours completing the questionnaire, reading and rereading it, correcting and explaining, adding and cutting. I imagined scores for every field and speculated on possible key words for the open-ended questions. I tried to be succinct yet thorough, sifting through my personal history so as not to forget anything – a course, an experience, some skill that might make a difference. From Monday to Friday, I spent long days in the classroom, listening to lectures on AM-X systems, followed by equally long evenings spent studying aircraft manuals and preparing for frequent tests. On weekends, I concentrated on the ESA questionnaire with maniacal attention to detail, an instinct picked up both in my engineering studies and in my military pilot's training. I was acutely aware that a detail on the questionnaire could spell the difference between entering the 'dogfight' – or instantly relinquishing my long-cherished dream.

My potential futures were now inextricably interwoven and would continue to be so throughout the coming year. Four weeks had gone by, and it was time to send in the questionnaire. Simultaneously, the initial series of lessons on AM-X systems was ending, making way for in-flight training. For the first time, and after much anticipation, I found myself at the controls of a proper combat aircraft, not just a training plane. The cockpit of an AM-X was where I most wanted to be while I waited to hear if I'd be going into space some day.

A couple of weeks later I was preparing for my first solo flight when the email I would anxiously await at every step of the selection process popped into my inbox for the first time:

From: astronautselection@esa.int
Subject: Congratulations!
Signed: Michel Tognini, Head of the European Astronaut Centre

I was overcome with joy and, more than that, a liberating sense of having escaped danger. It was only the first step, and I was one of a thousand remaining applicants, yet this meant that 7,500 people had already been eliminated. That stage was over, the one that had seemed to me most random, most unpredictable, most subject to chance. Now I was really in the 'dogfight'. My CV had aroused sufficient interest, and from now on I was convinced it would no longer be about formal requirements but about demonstrating the right combination of skills and personality. Even considering all my limitations and the ever-present possibility of making mistakes or meeting unfortunate circumstances outside my control, I now felt like I was in with a good chance.

Barely ten days later, I joined one of the first groups of forty applicants convened in Hamburg for an entire day of aptitude tests, a series of exams administered by computer, with brief intervals. The day was exhausting. The pace was fast, and the tests demanded a heightened level of attention and concentration for many hours. I found the tests in English, maths and mechanical comprehension something of a relief, since they had to do with acquired knowledge and therefore required less mental effort. The other tests, however, examined purely cognitive faculties such as sustained concentration, 3D visualization, visuomotor coordination, attention allocation and visual and auditory memory. We finished those tests with a sense of inadequacy, because you couldn't beat the computer: in each case, the speed and level of difficulty progressively increased until, inevitably, the human brain reached its limit, and the errors multiplied. Thanks to a good short-term memory and a certain familiarity with associative techniques for memorization, I was almost sure I hadn't made any mistakes on the test for visual memory. But with regard to the other tests, I shared the general sense of uncertainty. It was impossible to know how we'd done, and we could only speculate on the evaluation criteria. Only much later, while I was already in training with ESA, was I able to confirm one of the most widely held theories in the evening discussions back at the hotel: rather than achieving performance peaks on some tests, it was important not to drop below a certain threshold in any one of them. Today I know very well that an astronaut is not required to excel in any one particular thing yet should be able to manage fairly well in everything.

I left Hamburg feeling cautiously optimistic about my test results, and with mixed excitement and apprehension about other applicants I'd met. Many were engineers and scientists already professionally active in the space industry; some were actually already with ESA. It had been great to meet so many brilliant men and women, all of them passionate about space, to engage with them in enjoyable and stimulating conversations and discover a natural bond with hearts that beat with the same dream.

However, I also noticed how, having had my head down over the past few years, studying to be a pilot and taking my first steps towards a career as a military aviator, my connection with developments in space exploration had become increasingly tenuous. I was never obsessive about my dream to become an astronaut. I never collected autographs or souvenirs, didn't follow every little bit of news on the Space Station and I never tried to get in the loop about forthcoming astronaut selections. I'd always felt it was important to concentrate on the task of the day, and in recent years my daily routine had been grounded in my pilot training. But now, as I hoped to move forward in the selection process, I felt the need to immerse myself in the world of space. I had some revising to do and lots of gaps in my knowledge to fill.

On weekends I started going to a friendly café in Foggia with an internet connection where I caught up with several years of space news. If I went to the beach, I took along my books on space propulsion or orbital mechanics. In the car I listened to the audiobook of the first *Harry Potter* in Russian, a gift from a dear friend in St Petersburg. I chose a children's book in the hope that it would be easier and more suited to my rusty language skills. The most obvious result is that today I still have a small, but enviable vocabulary of Russian magical terms.

With hindsight and a long space mission under my belt, I am convinced that my experience in the Air Force taught me things no doctorate in engineering could have: discipline, humility, resilience, a sense of my own limitations, attention to detail, teamwork, leadership and also followership. I was confident that this sort of competence would make the difference in the end, when the highest-scoring applicants in the cognitive and psychological tests remained. But I also wanted to make sure that if I found myself among them, a question on Kepler's laws or an invitation to converse in Russian wouldn't catch me out.

At the same time, my AM-X course was very demanding, and the pressure of running two parallel marathons would lead to difficult moments in the months to come. Preparations for a flight of just over an hour required many hours of planning, and the in-flight training was accompanied by a packed programme of classroom lessons and tests, before we could move on to new and more complex types of missions. We had regular simulator sessions and thick training manuals to study the plane's most advanced systems, weapons and their use, communications procedures and rules of engagement. It was towards the end of that torrid summer in Apulia, when the only day off had been the public holiday of Ferragosto in mid-August, that the ESA email I'd yearned for finally arrived with the good news: I had been admitted to the next phase of the selection! There were now 192 applicants remaining.

We were invited in groups of six to the European Astronaut Centre in Cologne for a second day of psycho-cognitive evaluations. This time around, they focused not on individual performance but on relational skills, communication and problem-solving in groups.

I did the first test with Martin, a friendly German with a PhD in maths. The test involved a sort of video game, and our duty was to optimize car traffic. Each of us, individually, was responsible for half the city ('City West' or 'City East'), but the data necessary for making decisions were visible only to the other person, so we needed to exchange information rapidly and efficiently as well as give timely instructions to one another. In the few minutes we had to prepare, Martin and I agreed on what seemed like an effective protocol for communication, loosely inspired by the concise and effective radio jargon of military aviation. The initial warm-up run went quite well, it seemed, and we were pleased with our strategy, but to our surprise the observing psychologist was not impressed: she advised us to adopt a less severe tone for the actual test and to indulge in some polite formalities. Perhaps taking our cue from the dry and essential exchanges typical of military flight wasn't the ideal strategy for making a good impression. In the years to come, however, despite the short way in which we seem to have talked to each other, Martin and I would exchange emails regularly, jokingly calling each other 'City East' and 'City West'.

The group as a whole was later assigned a timed task to solve together. The problem was fairly complex, and we were being observed by a committee of psychologists, human resources professionals, doctors and a veteran astronaut. It's possible that there was no solution to the problem, but I couldn't help feeling some disappointment when, after a rather rambling discussion, our time was up and we had no suggestion to put forward. Whether for fear of exhibiting a tendency to dominate or reluctant to meet resistance from the rest of the group and end up in an embarrassing situation, no one had assumed the role of coordinator. I quickly weighed up the risks and forced myself to ignore them. My instinct was that this was not a moment for tactical manoeuvring. The only correct choice was to act in line with my character. As far back as I can remember, I've never been shy about assuming a leading role when necessary, and I wasn't going to start that day. When they presented us with the second problem, I leapt at the chance to put myself forward as moderator, and I made no bones about guiding the discussion, aiming to work in a structured manner so we could find a solution this time. Weren't we among the 200 applicants judged to be the best in an in-depth cognitive evaluation? When our time ran out, we had a valid proposal to present – and at the end of the test, the thanks I received from some of the applicants for taking the initiative convinced me that I'd done the right thing. I was pleased with myself, but that's rarely a good sign, and I'd have sorry confirmation of this during the interview each of us would have later with the observing committee.

It was late afternoon when my turn came, and I felt tired but confident. The applicants who'd already been interviewed had reported that the atmosphere was friendly and relaxed, and it seemed like the day as a whole was going well. So I took my place in front of the commission with a genuine smile. But when I stood to leave after forty minutes that seemed to go on for hours, my forced smile required all the discipline I had. I was sure that the selection had ended for me right there. A couple of members of the group didn't like me at all; I was convinced of it. The atmosphere was tense, and they asked me to account for my behaviour during the group exercise. It was clear that in their view I had been overbearing with the other applicants. Maybe, by nominating myself as discussion moderator, and in my zeal to obtain a result, I'd crowded out my colleagues. Maybe I'd failed to show consideration for all the ideas put forward. Or maybe I'd based my interaction once more on ways of communicating that were best suited to the military. Maybe, maybe, maybe . . .

In the weeks to come I would relive that interview hundreds of times, to the point where I hated myself for not being able to set it aside. It hadn't been only tension and awkward answers. A lot of it was relaxed and had gone well. I felt I'd made a good impression on several members of the committee, and it was even possible that the hostility I'd sensed from some of them hadn't been real, just a way of putting me to the test. But in any case, I was consumed by the agonizing fear that I'd let my dream evaporate in a matter of minutes, one afternoon at the beginning of October in Cologne.

The following two months of waiting were exhausting, probably the darkest period of my life. For weeks I lived with frequent headaches, and sometimes they only went away in the cockpit. With the exception of the Pill, I didn't have any other medication around at the time. Headaches were a rarity for me, and whenever I did have one, I calmly put up with it without taking painkillers. Even then, when they were plaguing me almost daily, it didn't occur to me to take anything for relief. My impractical approach to life had often been the subject of good-natured ribbing on the part of friends and colleagues.

In November, the course transferred to the Decimomannu air base in Sardinia for training at the nearby Capo Frasca firing range. This was a particularly demanding phase of the training, our commander advised us, and it was common to meet with difficulties. Sure as anything, I did, to the extent that they threatened the outcome of the course.

One Sunday I had news of an impending death in my family. I was plunged into a deep despondency. I cried at length and didn't even touch the materials needed to prepare for the sortie the next day, suddenly overcome by a paralysing sense of fatalism. That night, the AM-X course and the astronaut selection seemed only trivial details, and my interest in them a whim.

I've met many pilots with such natural talent that they can allow themselves to fly with little in the way of preparation or sleep, knowing they'll perform well anyway. I wasn't one of them. The right thing to do the following morning was to pull out of the flight. But I didn't. Due to a prolonged period of bad weather, we were already very late with the sorties planned for the course, and I'd lost extra days because of a nasty pulled muscle in my neck, so the thought of causing further delay felt unacceptable. I went up in the air as number 4 in the formation and, when circuiting the firing range in conditions of limited visibility, I created a potentially dangerous situation that cost the mission a negative evaluation.

Although it was a serious error, it was a fairly common one during training at the firing range, so the instructors weren't all that concerned. With the best of intentions, and to help me recover the situation as soon as possible, they added me to another formation going up soon after. Of course, no one could have known about the emotional state I'd been in the night before. Once again, I failed to bow out of the sortie, taking another irresponsible decision. Once again, I made a serious mistake. Once again, I got a negative evaluation. Two failed missions in one day. It's hard to imagine a worse situation: three negative evaluations in a row and I would have been expelled from the course.

With years of hindsight I can say that that miserable day proved to be a blessing in disguise. In a brutal way, it sent me a powerful warning I needed to hear: I could not allow myself to be distracted or to take decisions lightly. As is often the case, the threat of immediate danger unleashed all the power of concentration, and fortunately I continued to believe that I could get through if I gritted my teeth. There were also some lucky circumstances. As expected in these situations, I was placed under the care of a small group of particularly able instructors. And throughout those weeks of struggle, I could rely on the friendship of my three classmates, who were always there for me, extending a favour or just lightening the mood with a joke.

Wrapped up as I was in the course at the firing range, I gave no further thought to the interview in Cologne. My headaches went away and, as is so often the case when I'm utterly focused on a single objective, I began to feel well even though I knew there was still a steep climb ahead. I completed this phase of training at the firing range with no further hitches and I went back to the Amendola air base with a renewed, if cautious, sense of inner peace.

Late one afternoon, I was finalizing a map, using red circles to indicate the range of non-existent ground-to-air missiles scattered across a virtually hostile area my formation would fly over the next day, when I got an email that ended two months of enervating uncertainty.

> Dear Astronaut Applicant, it is my great pleasure to congratulate
> you . . .

Exultant, I left the room, bursting with joy and unable to contain myself. I'd waited for too long and thought too often that the outcome would be very different.

That evening, after pouring out my joy in emails to my family and friends, I joined ongoing conversations between applicants within the different groups that had sprung up after we'd met in Cologne and Hamburg. Many had received bad news and were leaving the selection process. We, the lucky ones, were going ahead, still incredulous, and we were at a loss for words with the unsuccessful applicants. We couldn't explain what had made us stand out and we felt their disappointment keenly.

We found out that there were forty-five of us left, and we were soon sifted into groups of seven or eight for a whole week of medical visits. I got a call one day from Brigitte, a French doctor I remembered meeting in Cologne. In order to schedule me on one of the weeks of visits, she needed to know the date of my last period. My potential employer's interest in my menstrual cycle definitely underlined – as if it were necessary – the fact that being an astronaut isn't any old job! As it happened, over the coming years, Brigitte's characteristic French accent would become very familiar as we continued our early habit of sharing personal issues. In fact, she was to become my flight surgeon. Not only did we discuss at length how menstruation can be suppressed in orbit, but she was the first person at ESA to learn, a few months after the mission, that I was expecting a baby.

Along with the summons, ESA sent a document briefly describing several dozen tests we would have to undergo. These were purely medical exams which ranged from a routine dermatological check-up to a bone-density check. There were no centrifuges, rotating chairs, isolation chambers or any of the other particularly arduous or spectacular exams the collective imagination associates with astronaut selection. There was therefore nothing for us to do in preparation, since you cannot hope to change your state of health in a few weeks. So I just carved out a bit more time for sports, tried to eat healthily and lost a few kilos in order to reach what I considered my ideal weight.

On the evening before the visits began, I had supper with some old university friends and then joined my fellow adventurers in a little hotel a few hundred metres from the entrance to the campus of the German Aerospace Centre. It was their Institute of Aerospace Medicine that would be conducting the medical evaluations. As I took my seat, I realized to my surprise that I knew the woman sitting next to me: Regina had begun the mechanical engineering course in Munich a year after me, and we'd lost touch with one another when we'd both gone abroad to study. Here we were ten years later, I, a military pilot, and she, a PhD in aerospace engineering with a demanding job in a consultancy firm. These days she is one of the happiest people I know, but Regina was then deemed unfit because of a completely asymptomatic medical problem. Her dreams were shattered because of a small detail completely beyond her control that had no bearing on her daily life. No doubt she would have been none the wiser if not for the tests imposed by the ESA selection.

The week flew by between medical exams and spells of waiting in a basement room filled with irreverent laughter and the sort of swaggering you get from military pilots, who made up the majority of our group. Cardiologists, orthopaedic specialists, ophthalmologists, ENT doctors, gynaecologists, psychiatrists, dentists, gastroenterologists, radiologists and ultrasound technicians . . . Over five days and according to a rigid schedule, we saw every type of specialist in an organized migration from one hospital department to another, one room at the Institute of Aerospace Medicine to another.

I left at the end of the week with some misgivings regarding my eyes, since after the first in-depth exam, which went on for several hours, I'd been subjected to a second check. I've always had perfect vision, and it seemed odd that my eyes might be letting me down, but the fact was, nobody else had been recalled by the ophthalmologist, though almost everyone had had to repeat at least one exam. There was no point in asking for an explanation: we would only be informed right away if they found a condition requiring immediate medical attention. Otherwise we would have to wait a few weeks to learn the results, as with all the previous steps in the selection process.

This time, I didn't have to wait for long. Around the middle of February I got the email I'd been waiting for:

From: astronautselection@esa.int
Subject: Congratulations!

So my eyes had not let me down in the end! The healthy body I'd won at the lottery of birth, which had never required much attention, which for so long never grumbled about being ill-treated by a hectic lifestyle, too little sleep, poor eating habits or psychological stress . . . the body that had only in the past few years begun to send me timid messages that got me to pay more attention to my lifestyle . . . that body was still giving me its full support: you're off to space, Samantha!

One small detail remained: the selection process was not yet over. Before the end of the day, emails began to fly thick and fast among the applicants: some had already been contacted, some would hear the good news or the bad only the next day, after twenty-four hours of torment. Our pan-European intelligence network soon revealed that there were twenty-two of us still in the running. Despite our best attempts, it was difficult to calculate our chances for success. We didn't even know how many of us would be chosen. Four? Five? Six? ESA had never stipulated an exact number. Surely there would be some sort of balance between nationalities: it was unthinkable, for example, that two out of five would be British, or three out of six German. We knew that if the selection process to this point had been solely based on single-applicant evaluations, from now on the balance and composition of the final group would be the main objective. All of us who had got through to this stage were probably suitable as astronauts. Now ESA would have to put together a team.

Of course, there's always a way to ruin a good impression: showing up poorly prepared for an interview with ESA's senior managers, for example. That would be the final stage of selection, or so we thought; we still didn't know about the additional interview with the director general. It wasn't the time to let down your guard; quite the contrary. Summoning up all my courage – or maybe my *chutzpah* – I knocked on the door of the squadron commander, the officer ultimately responsible for completing our training. He was a reasonable man and, as it turned out, really passionate about space. What's more, he had always shown himself to be agreeable towards me, something I didn't take for granted. I was, after all, hoping to be selected as an astronaut, and that would have made my AM-X training irrelevant for my future employment. I was also asking a great deal. I wanted a week off from training to focus entirely on preparing for the coming interview. It was really bad timing for such a request, since our course had already been extended beyond its usual length. But I had to ask. Twenty-two applicants remaining. It seemed all I had to do was reach out my hand to touch my dream.

Thankfully, and to my immense surprise and relief, the squadron commander agreed without a moment's hesitation. Years later – he was then a friend – he came to Baikonur to attend my launch and a few weeks later he wrote, 'You seem to have been born to do what you're doing.' Who knows? Maybe he was thinking that in his office in Amendola, when we had that unusual conversation.

For one week, I forgot all about the AM-X in order to immerse myself totally in the world of spaceflight. I even spent one Saturday shopping, an activity I find about as appealing as a trip to the dentist. I explained to three sympathetic assistants in a shop in the centre of Foggia that I was looking for an outfit for an important interview, and they spent a lot of time and took great care in helping me choose a simple, yet elegant suit. And then I set off for rainy Noordwijk, a Dutch city on the cold North Sea and the site of ESTEC, ESA's European Space Research and Technology Centre.

The interview turned out to be focused on personality rather than technical skills. In a relaxed, occasionally even cheerful atmosphere, committee members took it in turn to ask questions, many of them about how I would behave in situations of interpersonal conflict, where there was disagreement over values and priorities or disappointed expectations. Apart from the line-up of senior managers – they'd hardly attend an ordinary recruitment of young applicants, I thought – none of it seemed very different from a normal job interview.

I left the interview calm and confident. I felt we had connected and it didn't seem that any problems had come up. As far as I knew then, the selection process was finished. I'd done everything in my power, given it my all, and it was no longer in my hands. I couldn't do anything now but wait. I had no idea as yet how the long wait ahead would wear me down, nor could I anticipate how exhausting those last nerve-racking days would be, when an announcement was imminent, and I jumped nervously every time the phone rang, at every flash indicating a missed call. My emotional investment was now so great that a negative outcome would have left me staggered.

So on that May evening in my little room at the Istrana air base, I welcomed the news of my selection with an immediate release of tension and a profound sense of relief. Only in the coming days did trickles of joy, satisfaction, even a little pride about joining the European astronaut corps, slowly soak every fibre of my being.

3.

> I was sitting in a chair in the patent office at Bern when all
> of a sudden a thought occurred to me: 'If a person falls free-
> ly he will not feel his own weight.'
> Albert Einstein, Kyoto Conference, 1922

Kennedy Space Center, Florida, 8 February 2010

'T-9 minutes and counting,' a voice comes over the loudspeaker.
It's only just past four in the morning, and the illuminated panel
in front of the stands indicates that the countdown has started
again after a planned hold. In front of us stretches a mirrored expanse of water, barely ruffled by a light breeze, but all eyes are
trained beyond it, about 6 kilometres away, where powerful spot-
lights are washing over Launch Complex 39A. There, rising from
the launchpad in all its glory, is the Space Shuttle Endeavour, and
in its capacious cargo bay are two new modules for the International Space Station: Node 3 and the Cupola. They arrived last
June from Turin and will go to complete that fabulous human
outpost in space, continuously inhabited by astronauts for almost
ten years. The Cupola, with its big windows, will offer them a
breathtaking view of the Earth.

I look up somewhat worriedly at the sky, where starless patches betray the presence of clouds. I'm quivering with the desire to
see my first space launch, but it may not take place tonight. The
first attempt, yesterday, was postponed because of the clouds,
and the odds of launching today are estimated to be 60 per cent.
The Space Shuttle is an incredible machine, but also an extremely complex and demanding one. It's not only the weather conditions here at Kennedy Space Center that must be favourable, but

also those in at least one of the three emergency landing sites on the other side of the Atlantic. Knowing full well that a launch can be repeatedly delayed, I've given myself a good two-week margin. I have no intention of missing this launch, since I won't have another chance to come to Florida during my basic astronaut training, and the Space Shuttle will complete its last mission in a few months.

My being here today is actually a very lucky circumstance. Right now, my colleagues are in St Petersburg for an intensive Russian course, which wasn't considered necessary for me. After the first month of lessons in Germany, I asked to take the exam to certify my proficiency level, which was deemed more than sufficient to fly in a Russian space vehicle. As much as I cherish the dream of improving to the point where I can enjoy Dostoevsky in the original one day, a holiday in Florida for the launch of Endeavour was more enticing than a course in St Petersburg in February.

The Space Shuttle has been a constant presence in my world. I was four years old when Columbia made its maiden flight, and I well remember the pictures of the Challenger accident on television when I was nine. I was a teenager when, for the first time, a fellow Italian went into space, making it seem all the more possible that, one lucky day, it might be me. And it didn't escape my attention when, a few years later, Colonel Eileen Collins became the first woman to serve as pilot, and then commander of the Space Shuttle. Unfortunately, my generation of astronauts won't be able to fly in this exceptional vehicle before it's retired from service. Until new American space capsules are developed, the only way of reaching the Space Station will be to fly in the Russian rocket Soyuz, and in the small spaceship of the same name.

The countdown resumes after a planned hold. Seven minutes from lift-off, the access bridge that allows the team to take their seats on Endeavour is retracted. The large external tank, bright orange, contains 800 tonnes of cryogenic propellants, liquid hydrogen and oxygen at very low temperatures. A small amount inevitably evaporates and is discharged through dedicated relief valves, creating the characteristic puffs of white vapour that shines in the lights on the launchpad. When the main engines are switched on, powerful turbopumps will funnel hydrogen and oxygen into the combustion chamber. But it's the solid propellant boosters – those two huge, white, cigars on the sides of the external tank – that provide the first and most powerful push against Earth gravity.

I wonder what the six crew members are thinking right now, seated uncomfortably for hours in their orange spacesuits, their backs parallel with the ground. It's the first mission for one of them, pilot Terry Virts. I can imagine he's consumed with the desire to lift off.

At T-4 seconds, I watch as the main engines are lit. Lift-off isn't yet inevitable: if one of the engines should fail to work properly, there would still be a way to halt the launch sequence. But not once the solid-propellant boosters are ignited as the countdown comes to an end and Endeavour, heedless of its 2,000-tonne weight, lifts off over Cape Canaveral in a sudden explosion of light so bright it turns night to day. Despite the distance, the noise is intense, and my whole body begins to vibrate with the sound waves. Enveloped by white vapour from the cooling water, the launchpad continues to radiate an intense glow, while the Shuttle climbs rapidly through the sky towards the northeast. It ascends through thin layers of cloud which, with the approach of Endeavour and its dazzling tail, emerge from the darkness in a play of light and shade worthy of Caravaggio – and are then swallowed up again by the blackness of night.

The boosters shut down after two minutes, as planned, and then drop off into the ocean, where they will be retrieved for refurbishment. The main engines continue to thrust Endeavour on in its race towards orbit. For nearly seven minutes I manage to follow it, a point that grows increasingly distant and imperceptible, until it disappears completely behind high and distant clouds. I couldn't have wished for a more spectacular launch.

Fifteen years ago, when I was an exchange student in the United States for a year, I spent a week at Space Camp in Huntsville,

Alabama, not far from where I am now. We studied the Space Shuttle for fun, as much as we could in the short time available, and then we simulated a twenty-four-hour mission. And now, thinking of the Endeavour crew about to reach orbit, the images overlap with my memories of that experience.

I can't imagine any better occasion than this spectacular launch for properly taking leave of that teenage Samantha. The time has come to move from the dream of being an astronaut – a brilliant, but vague dream – to the concrete reality of building an astronaut's future. Or maybe this transition is just an illusion: maybe the beating heart of the child who yearned for space, maybe the dreamy gaze of the girl who lost herself in sci-fi adventures are still there. Maybe these aren't only faint traces, but remain the mysterious source of my actions and feelings.

All of them – the child, the girl, the woman – are alive within me, the apprentice astronaut.

—

I couldn't know that night that, only a few years from then, my own life would be inextricably intertwined with the life of the Space Shuttle pilot then making his first space mission. Nor could I have dreamed that the man standing next to me on the bleachers, whom I had only just been getting to know, would become my life's companion.

For now, looking ahead, I could only see one certainty: the basic astronaut training, which, for the first time, was taking place in Europe. At the press conference in Paris the previous May, I had had the first opportunity to meet all together my five companions in this venture in one place. It's where I got to know Tim, Alex and Thomas and met up again with Andy and Luca, whom I'd originally got to know during the medical exams. We were more or less the same age, and we were going to spend the fifteen months of the course together. Luca and I shared not only Italian nationality; we also both belonged to the Air Force. He'd joined up when he was much younger – right after high school, and during those months he was finishing the course as a test pilot for rotary-winged aircraft. Together we took part in another press conference the next day, in the crowded press room of Palazzo Chigi, seat of the Italian government.

In the weeks that followed, the Air Force received an unusual number of requests for interviews and public appearances. Their decision, at least initially, was to accommodate the media interest and try to satisfy it. It embarrassed me, but surprised no one, that attention was primarily focused on me, 'the first Italian female astronaut'. While I found myself fielding questions that were often frivolous, along with expressions of admiration and astonishment that clashed with my sense of reality, Luca took things with his customary self-deprecating sense of humour. He'd often smile and introduce himself to journalists with the greeting, 'Hi, I'm the other one!'

We hadn't had any kind of press training, but one day ESA sent us some guidelines. I still remember four of them because even now I consider them the Golden Rules: never lose your cool; speak only about what you know; don't discuss politics; don't discuss your family or your private life.

They were simple rules based on common sense, but they were useful in terms of steering through that early summer maelstrom. At the time, I didn't know yet how to separate myself from my media image or the public conversation of which I was the subject. And I wasn't free to decide, to select or to request appropriate forms of collaboration as I am these days, most of the time. So I was very happy to use my saved holiday in order to retreat from the bewildering clamour of those weeks.

In view of my new ESA assignment, I needed to find accommodation, so I went to Cologne, where, for the first time in my life, I signed a rental contract for something more than just a furnished room. I chose a small apartment in Sülz, a pretty area on the left side of the Rhine only fifteen minutes by tram to the centre of the city and its famous Gothic cathedral. I moved in late one evening a few weeks later, having driven all the way from Italy. I had with me a few suitcases, a chair and a camping table, and my computer and printer. I arranged those few things in the living room, the only room with a working light bulb, and blew up an air mattress to sleep on.

Happiness always takes me by surprise, sneaking in unexpectedly, almost inexplicably. I think it must grow inside, slowly and quietly, until some small detail makes it spill over, and all at once it saturates body and soul. Or maybe it's always there under your skin, waiting for you to make room for it. That night, I went to bed coddled by the expanse of clear sky I could see through the big windows in my living room. The full moon came out, haloed in silver – and suddenly I was happy.

On 1 September, we showed up at the European Astronaut Centre (EAC) for the first day of school, curious and excited about our future training. I had visited EAC more than ten years earlier, when I was still studying aerospace engineering. The opportunity had been offered during the course on human spaceflight taught by the German astronaut Reinhold Ewald. I still have a photocopy of a cover article from a 1998 issue of *Scientific American* that Reinhold handed out to us at the time, and in which the American astronaut Shannon Lucid talked about her extraordinary experience: 188 days on the Russian Space Station MIR during the pioneering years of space collaboration between Russia and the United States.

I discovered that Reinhold's office was now right across from mine at EAC. Or rather, ours, since I would share my office with Andy, a Danish aerospace engineer with a PhD and some years of experience working on offshore oil rigs in West Africa.

We wouldn't be spending much time in our office over the fifteen months of basic training. From the very first day, we had a full schedule that usually kept us busy from 9 a.m. to 6 p.m. The initial four weeks were spent mostly on introductory lessons, some of them technical and many organizational: How does ESA function? What are its main programmes? How is astronaut training organized? In a classroom reserved for us, we watched thousands of PowerPoint slides. Some days our schedule included sports; on others we had to carve out time at lunch or the end of the day.

Although most lessons were not that demanding, the days were still very long. I sometimes managed to leave in time for a quick trip to IKEA on my route home before it closed. But there were days when I got home late to my apartment, where, in addition to furniture, I still lacked internet service, a washing machine and curtains.

Despite this, I immediately felt at home in Cologne, a welcoming city that doesn't take itself too seriously, so much so that Carnival season there goes on for three months.

The excitement of a spaceflight was some time away from those early days of sitting in lessons, but I continued to feel like Alice in Wonderland. EAC was a special place, a small centre with roughly 100 specialists from all over Europe, including Russia, who were dedicated to the support, training and medical care of astronauts. To my delight, I rediscovered the beauty of a multicultural environment – starting with the pleasure of hearing conversations in many different languages. The lessons, too, were becoming more interesting. A series of short, intensive courses began on various subjects – from biology to geology, information networks to crystal growth – each with a short final exam. The aim was to provide us with a kernel of knowledge in areas related to the technological and scientific aspects of spaceflight, so that we'd be able to communicate effectively with the scientists and technicians we'd be working with once we were assigned to a mission on the ISS, or International Space Station.

We knew very well that our finish line, the ISS, was still a long way off. In the best-case scenario, the first of us would go to space four years from then, and at worst, the last of us might have to wait for up to ten. We were six able and ambitious apprentice astronauts, and I had no doubt that all of us would complete the training with excellent results. The order in which we went into space would depend on the rotation of missions among the ESA member states and not on our proficiency. In a sense, this was a relief. We weren't in competition and could dedicate ourselves unreservedly to the success of our group as a team.

We spent those first months in the little bubble of our raised classroom, which overlooked the large training hall dominated by a mock-up of Columbus, the European laboratory of the ISS. The mock-up was a full-scale replica of the module that is actually in orbit, a sort of large aluminium can measuring 4.5 metres in diameter and similar to many others, longer or shorter, that make up the Space Station. Each one can be an independent pressure vessel, with hull and hatches capable of maintaining an internal pressure equal to that found on Earth at sea level. In fact, the modules are linked and hatches are usually kept open to create a single internal volume of 930 cubic metres, similar to that of a large airliner. Sure, it isn't completely habitable: the 'furniture' occupies much of the space. In the case of the ISS and specifically the non-Russian modules, this furniture is composed of 'racks', units similar to tall, narrow wardrobes, with a flat front, but a curved back that follows the cylindrical shape of the hull. These racks are distributed along the entire length of the modules, one next to another in such a way as to create an internal cabin with a square section that has 'walls', a 'floor' and a 'ceiling'. Even with the racks, there's a lot of free space: I could stand in the middle of the Columbus mock-up without being able to reach the side walls.

We apprentice astronauts waited with some trepidation for the moment when we could begin our lessons on Columbus. Immersed as we were in courses on the basics – and even as we enjoyed the subtle pleasure that comes from learning how things work, whether natural phenomena or products of human ingenuity – we felt we hadn't yet entered the real world of training, the world inhabited by the astronauts assigned to the next Space Station crews – American, Russian, Canadian, Japanese or European colleagues we caught sight of every now and again from our classroom windows as they went to their classes in the mock-up. We couldn't wait to leave our desks for the training hall, or the deep pool where we would later be taking a useful preparatory course on extravehicular activities, or EVAs. Often called 'spacewalks', a term that doesn't do justice to the effort they require, these extravehicular activities are practised underwater, where it is possible to simulate some aspects of weightlessness, the condition of those who do not feel the effects of gravity.

We know it's not possible to free ourselves from the force of gravity itself – certainly not on the surface of the Earth, nor in the Space Station, which orbits at an altitude of only 400 kilometres, a short distance that weakens the Earth's gravitational pull by only about 10 per cent. Besides, gravity – at least as far as we've been able to observe, and according to our mathematical models – is only a force of attraction, without an equivalent force of repulsion to balance it. In other words, it cannot be shielded.

However, before our basic training ended, we'd have the experience of a few minutes of weightlessness under our belts: during a parabolic flight, we'd find ourselves floating inside an aeroplane cabin just like astronauts float around in the Space Station. In both, weightlessness is due to the condition of freefall, as in Albert Einstein's famous lift: the occupant becomes weightless when the cable is cut, since she is falling with the same acceleration as the lift itself. If she were standing on the scales, they would inevitably register a weight equal to zero.

So if the ISS astronauts are in freefall, why don't they ever crash onto Earth's surface? Because they are moving at the dizzying speed of 28,000 kilometres per hour, or about 8 kilometres per second; their trajectory is constantly bent by the force of gravity of our planet, yes, but only just enough for it to follow the Earth's curvature. They never fall to Earth, because they are condemned by their great speed to miss it; they are constantly falling *around* it. Just like the Moon, which is a thousand times farther away, but also compelled by Earth's gravity to wander eternally around our planet.

4.

Bordeaux, 7 May 2010

It's the most eagerly awaited day of basic training. Today we'll get to try absolute lightness for the first time. For short intervals we'll be free of the weight that otherwise accompanies us every moment of our terrestrial life. Like astronauts in orbit, we'll float in freefall. For twenty-two seconds at a time, to be exact, twenty-two seconds in which we'll be hurled through the sky like helpless stones and we'll be following a parabolic trajectory, just as any self-respecting stone does or the jets of water in a fountain, unless they just happen to be thrown up vertically, allowing no arching trajectory. But we're not concerned with that right now, since we're in an Airbus A300, which would never even think of ascending vertically. It will pull up its nose to 47 degrees, stopping somewhere roughly halfway between a horizontal flight and a theoretical vertical ascent. From there, the pilots will begin to cancel out all aerodynamic forces so as to trace a parabola through the sky, like a stone. And inside the cabin, we'll be forced by the simple laws of gravity to follow the same trajectory, with the result that we'll find ourselves floating.

We've taken off, and there's a palpable excitement in the air, even though we've all been injected with an anti-nausea drug that makes us a little drowsy. The few rows of seats in the cabin are at the far front of the plane; we have to remain seated until we get to the designated area over the Atlantic Ocean. Tracing parabolas in the sky, each time climbing more than 2,000 metres in altitude – as if riding on a gigantic, invisible rollercoaster – is definitely not permitted in everyone's airspace.

The rest of the cabin is fitted with protective padding and divided into stations where we will perform exercises in dexterity and the positioning of our bodies in space, while the plane traces thirty-one parabolas. Nothing difficult, just short tests to familiarize ourselves. We'll throw each other large cubes of different mass, for example, to see that we need to find a solid anchoring point in order to avoid being pushed backwards in reaction to the throw. We'll practise rotating our bodies in three dimensions, using only our hands and the handrails, as we would during a spacewalk. We'll even be able to run, our legs lighter than those of a marathon runner. I saw the treadmill when I came in, mounted vertically on the wall just like the one in the Space Station. There's a bench on the floor that you can lie down on, with your feet resting on the running platform while you wait for the parabola to begin. You may wonder how it's possible to run in weightlessness: don't we bounce off at the first step? Definitely. That's why we have to wear a harness, anchored at both sides of the platform with two elastic straps.

At last we get permission to stand up and, without further delay, we start moving around the empty part of the cabin. We want to be ready for the first parabola. It's all ours: twenty-two seconds of total freedom to explore weightlessness. We'll start our exercises on the next one. Our faces light up with the smiles of excited children. As adults, we don't often get to experience completely unknown sensations: the primordial pleasure, both physical and mental, of discovering new ways of inhabiting the space around us, new ways of relating to it and moving through it, is a distant memory from early childhood. And here we are, more than thirty years old, anticipating that pleasure.

The aeroplane points up, and I feel pressed towards the floor, my apparent weight double the norm: it's the 2-G pull-up that starts the manoeuvre, or 1.8 G to be precise. I follow the recommendations we received yesterday during the pre-flight briefing, keeping my head still and looking ahead. I may have taken anti-nausea medicine, but I don't want to tempt fate and risk ruining this experience. We keep pulling up. A voice from the cockpit calls out a sort of countdown in attitude degrees: 'Thirty! . . . Forty! . . . Injection!'

It's a gentle transition: all at once, I simply notice that there's nothing keeping me tethered to the floor of the plane any more. I give myself an almost imperceptible push and find myself floating through the air, free of my own weight – as in my recurrent dreams, which are sadly becoming less and less frequent as I age. Or like on a scuba-dive, hovering over the seabed but without the encumbrance of all the gear or the viscous heaviness of the water. I bump against the ceiling awkwardly, try to keep my bearings and attempt to dodge my colleagues' knees or elbows, exchanging good-natured shoves with them as we collide and sending each other flying towards opposite sides of the cabin. Someone takes advantage of the crew to help them curl into a ball and rotate in the air. Someone else acts like Superman.

As we playfully float through the cabin, the plane reaches the apex of the parabola and then dives into its descent. Twenty-two seconds go by quickly, and soon enough we hear that voice again through the loudspeakers: 'Twenty!' We're already twenty degrees nose down. We all hurry back down to the floor of the plane, grabbing the straps with our hands and feet. 'Thirty!' The crew members try to assist anyone who's clearly dawdling. 'Pull out!' Another 2-G pull-up, as the pilots bring the airplane out of the dive. My body regains weight and substance, and I'm pressed hard towards the floor again. Someone next to me has been caught out in mid-air and falls with a thud. In the confusion, some remain lying on the floor, and others stand, waiting for the pull-out manoeuvre to end. I keep my head still, only my eyes moving as I search the others' faces, creased with innocent joy. We exchange smiles of mutual delight.

Shedding your own weight is an explosion of freedom.

—

The parabolic flight in Bordeaux was part of a short European tour of locations related to ISS operations, and the itinerary seemed designed to plunge me into memories of my university days. One of our stops was in Toulouse, where I'd been thanks to the EU Erasmus exchange programme; for a few months I'd worked on an experimental thesis at SUPAERO, one of the French *Grandes Écoles* of engineering. Perhaps I spent too much time in the propulsion lab taking measurements on a turbofan, because I never met my future colleague and fellow astronaut Thomas, who was a student at SUPAERO at the same time. After working as an engineer for a few years, he had, like me, pursued his passion for flying and become an airline pilot.

What took us to Toulouse was the ATV-CC, the Control Centre for the European cargo vehicle ATV (Automated Transfer Vehicle). In the preceding weeks of basic training, we'd studied its systems, in particular the rendezvous and docking with the Space Station. The ATV's special feature was in fact its ability to perform a completely automatic approach and docking, a technology considered so reliable that the possibility of remote manual piloting from inside the ISS was not envisaged as it was with the Russian cargo vehicle Progress. There was, however, a final level of safety ensured by human beings. In the years of preparation for my mission on the ISS, during dozens of hours in the simulator, I would be trained to visually estimate the position, distance and attitude of the ATV, using graduated scales manually superimposed on the video camera images – a technique somewhat less than futuristic but undoubtedly handy and reliable. Human spaceflight is, after all, a mixture of old and new, of sophisticated technology and creative tinkering, and perhaps nothing illustrates this better than the quaint image of astronauts looking at black-and-white images against a laminated scale, intent on identifying any anomalous parameters that might raise fears of a collision, and prepared to press a large yellow or red button if necessary.

Our European tour also included a stop in Bavaria at the Columbus Control Centre, or Col-CC. Located in Oberpfaffenhofen, in the picturesque region of lakes and forests west of Munich, Col-CC responds to the callsign 'Munich' when it's called over the radio by the Space Station crew and is responsible for the systems of the European Columbus Laboratory and all the ESA experiments carried out on board the ISS. The visit to Col-CC was our first real chance to make contact with the world of operations, or the daily management of on-board activities in real time. There was much affection for, and a certain curiosity about, our small group of apprentice astronauts, and we responded, keen to get to know the ISS community beyond our small EAC.

By now, we'd bonded as a team and had even jokingly given ourselves an unofficial class name, the Shenanigans, inspired by our propensity for pranks. It may be that we'd become a little restless. To be sure, the training was now less theoretical and more demanding, but I think we all felt that we could do more and learn at a faster pace. Our visit to Col-CC of course didn't mean we were about to set forth on our first spaceflight, but it made us feel a little closer to the reality of actual missions on the ISS. For me there was also a truly memorable moment when I could talk to the ISS crew for the first time on Space-to-Ground, even if it was just a simple courtesy greeting on behalf of the Shenanigans.

We left Munich the next day, but not before sharing a typical Bavarian dinner with the staff of Col-CC. In the midst of all the amusement, we were encouraged to participate in lively traditional dances in local costumes. Voluntarily, of course. It's well known that an occasional willingness to make yourself look ridiculous is a solid basis for team spirit – especially when you're new to a group.

We travelled a lot over the coming months. First there were short visits to Houston and Tsukuba, where we would one day be training on the American and Japanese components of the Space Station. Then we went on a survival course in the atmospheric Sardinian highlands. Whenever we were at EAC, we spent our days in the pool or the training hall, either in the robotics simulator or the mockups of the ATV and Columbus, which included pretty high-fidelity models of the European scientific racks for biology, fluid dynamics and human physiology. We had finally begun to prepare for the two most distinctive astronaut activities: performing spacewalks and flying the robotic arm on the ISS. In both cases, our basic training offered introductory courses, the first building blocks upon which, over the years, we would further build and consolidate our skills.

The full name of the ISS's robotic arm is Space Station Remote Manipulator System, but as with almost everything in the space community, it's usually referred to by an acronym: SSRMS. Since it's built in Canada, it's sometimes called Canadarm 2 in order to differentiate it from the original Canadarm, the robotic arm on the Space Shuttle. Have you ever wondered how it was possible, in space, to assemble a structure as large as a football field and weighing more than 400 tonnes? It was by using the SSRMS that modules, radiators, solar panels and the truss – the long external scaffolding on which these are mounted – were joined together. You can imagine it as an imitation human arm, with two cylindrical booms about 8 metres long and three joints: the shoulder, the elbow and the wrist. They're actually called by those names, and they have the same degrees of freedom as human joints. Let's take the shoulder, for example: if we hold an arm straight out in front of us, we can raise it and lower it, move it from right to left and vice versa or rotate it, as if we wanted to twist a doorknob while keeping our wrist still. It's easy to see that your wrist has the same three degrees of freedom, while the elbow has only one, allowing it to bend the forearm towards the shoulder.

Some years and hundreds of simulator-hours later, I would find myself at the controls of the real SSRMS, one of the most intense moments of my space mission. For the time being, at EAC, I had several appointments per week with Boris, a simulated robotic arm resembling the SSRMS but located in an environment that was much less complicated compared to outside the Space Station. Boris existed only as a simulation, as if in a video game, and yet the two hand controllers – the joysticks, let's say – were entirely real, and exactly like those of the robotic workstations on the ISS. The right one controlled rotations and the left one linear translations. One of the skills we had to acquire, and on which we were continually evaluated, was the ability to deflect the controllers smoothly and progressively. In fact, starting or stopping Boris's movements without being cautious immediately caused undesirable oscillations that were hard to control. In addition, as when piloting an aeroplane, it was necessary to develop a so-called scan pattern, an automatic routine to monitor a series of relevant parameters in a continuous cycle: Is the arm moving correctly? Is it far enough away from the structure? Am I moving the joint to its limit or causing two of the robotic arm's components to collide with each other? Is it possible that I'm getting close to a singularity, one of those joint positions in which the mathematical model fails and the arm locks up and won't move? Are the video cameras still working well? Do they need to be redirected or should different ones be chosen to follow the current movements? All of these are simple exercises on their own. And yet, as with piloting a plane, the difficulty lies not in each individual task, but in repeating an entire series of actions quickly and continuously, and never focusing attention on just one of them. During our sessions with the simulator, any distractions cost us instantly, since our instructors made sure that collisions and singularities were always lying in wait for us.

At the end of our first summer as apprentice astronauts, we passed the introductory exam in robotics and left for the last stop of our basic training: the legendary Star City.

5.

> But Sasha . . . was from Russia where the sunsets are longer, the dawns less sudden, and sentences often left unfinished from doubt as to how best to end them.
> Virginia Woolf, *Orlando*

Star City, 12 September 2010

It's a moment I've been looking forward to: at the end of our basic training we're about to start a two-month course on the Russian segment of the Space Station and an introduction to operations on the Soyuz vehicle. Nikolai, a man with very white hair and a sweet smile, has come to collect us from the airport in Domodedovo and, having struggled through the slow and messy traffic on the ring road around Moscow, he drives the little ESA van onto less congested roads flanked by birch and fir trees. After more than two hours of travelling, a block of white stone decorated with a large four-pointed star appears at the side of the road. On top, carved in relief, sits the word *Zvezdnyj*. We're close by now: nestling, almost hidden in these majestic woods 25 kilometres or so north of Moscow is Zvezdny Gorodok, Star City. The place is legendary, a site where the history of humans in space has been written for half a century. Once a secret military citadel, today it's a hub in the web of international collaboration that keeps the ISS going. It's here that Yuri Gagarin and Valentina Tereshkova prepared, here that Space Station crews train today for everything concerning Russian components and, naturally, the Soyuz vehicle. At moments like these, I still feel like I'm living in a fairy tale. Only nine years ago, I left a Moscow that was much less rich and modern, with my dissertation in my computer and, in my pocket, the summons to the Air Force Academy for a short internship, the final phase in the admissions process.

Our van slows down, bringing me back to the present. The road ends in a large open area walled on two sides. We've stopped in front of the only open crossing, a checkpoint manned by two lazy guards in military uniform. They exchange a few words with Nikolai in that brusque, dry tone which isn't considered rude in Russia, and, having verified his pass, they move aside to permit access. We drive down a straight and narrow road with trees on one side and a stone wall on the other. Many years of reading science fiction make me feel like I'm travelling down a tunnel into another dimension. Behind us, the frenetic twenty-first century in Moscow; ahead, a mythical place that seems protected and suspended in time. There was no arrogance in taking away space from nature: there's still forest everywhere and in the small lake we see reflected the bright colours of what the Russians call 'golden autumn'. There's nothing to make you think that any significant changes have taken place since the time when you could easily have bumped into the young pioneers of space exploration criss-crossing the woods on cross-country skis. Chaika – Tereshkova's callsign on Vostok 6 – never actually left. Though she's become an important public figure, actually a member of the Duma, she still has a house here, and it's said she often attends the parties and celebrations held by the cosmonaut community.* Gagarin, who died a few years after his historic flight, is also present in his own way. His stern statue guards the access road: in typical Soviet style, he's portrayed not in his cosmonaut suit but in his ordinary work clothes, lightly outlined, holding a flower in one hand behind his back.

I am increasingly impatient to explore this place I've so often imagined, but right now there's a more immediate issue: we're late for supper. As it happens, Scott Kelly and Ron Garan, two

* In accordance with convention, Russian and Soviet astronauts are referred to in this work as cosmonauts.

NASA astronauts whom I got to know a few weeks ago at the EAC, have offered to have a bite to eat ready and are waiting for our arrival. So we quickly leave our baggage in our simple but comfortable rooms and we're assigned bicycles that, starting tomorrow, will be our means of local transport. We then ask Nikolai to drive us to Cottage 3, where we are expected.

We've all heard about these cottages: a bizarre architectural witness to the American presence here since the 1990s, when several NASA astronauts flew on the Russian Space Station MIR. Most of the residents of Star City live in poky apartments situated in austere and imposing concrete blocks from the Soviet era, their windows tiny and their balconies merely decorative. European astronauts are housed on the second floor of the so-called Profilaktorium, a reasonably welcoming structure where cosmonauts spend a period of quarantine post-flight. NASA astronauts, on the other hand, live in three two-storey wooden cottages painted white. Between a characterless, unfinished residential block and a recent Orthodox church painted bright blue sits a little piece of North America – complete with waste disposals in the sinks and 110V electrical plugs.

Our supper is cheerful and informal. The others have already eaten, and we're invited to help ourselves from pots kept warm on the hob. Excited and in good spirits, we try to participate in the conversation, interjecting a clever line here and there in hopes of eliciting a laugh. But mostly we listen, fascinated. There are astronauts currently in training here, the doctor stationed in Star City and various NASA support staff. Some of the exchanges are incomprehensible to us: there are too many terms and acronyms that we don't know as yet, too many names of people we've not yet met. Now and again we ask a question, and now and again someone offers an explanation. Those who've already eaten and feel they've made an honest contribution by washing up or helping to clean the kitchen are disappearing into the basement. We're invited too. We go down a little stairway and find ourselves in Shep's Bar.

'Shep' is Bill Shepherd, originally a US Navy Seal, then a NASA astronaut, and finally commander of the first expedition on the ISS. Ten years ago, he, Yuri Gidzenko and Sergei Krikalev reached an embryonic ISS, then composed of only three modules, and from that time on there has never been a single day when humans have not been in space. To prepare for that historic mission, which was postponed several times, Shep spent long periods of time here in Star City, and one day he had the idea of transforming the basement of Cottage 3 into a place to relax and socialize. In addition to the bar, a battered sofa, a ping-pong table, a pool table and a huge television, there are dozens of objects and knick-knacks I imagine have accumulated over the years: pennants, trophies, photos, and autographs of people who've passed through, some more famous than others . . . Moscow's nightlife is one of the most vibrant in the world, but perhaps the most exclusive venue of all is found in a humble basement in Star City.

The evening ends early, and we walk back towards our rooms. The Profilaktorium is very close to the cottages, just on the other side of a wide court, but because there are five of us here at the same time, we're in a sort of guesthouse about a ten-minute walk away. It's a pleasant stroll on dirt paths beside the lake. We don't meet a soul, other than a few stray dogs, and we have to use our pocket torches as there are no roadside lamps and the Moon has already set. My impressions of the day are settling down in me, and all around reign calm and silence, ideal companions for savouring this evening's happiness.

—

It's easy to explain why our small group of apprentice astronauts was so enthusiastic about Star City and its warm welcome. After nearly a year of basic training, we had acquired a great deal of knowledge and skills, but we'd had little contact with the international human spaceflight community, and especially with non-European astronauts. It's true that in the spring of that year we'd spent a week at the Johnson Space Center (JSC) in Houston, but our trip had been too brief and our activities too isolated for us to feel more than short-term visitors. We had only brushed up against that complex world, bustling with activity, and it had barely registered our existence.

We had, however, begun to forge friendly links with our parallel class, the Chumps, i.e. the American, Canadian and Japanese astronauts, who, like us, had been recruited the previous year and were now busy with their basic training at JSC. These links were strengthened when they came to Cologne later that year, thanks to a now-legendary party at Andy's house.

That evening in late July, we uncorked a bottle of *spumante* to celebrate Luca's assignment on Space Station Expedition 36/37, with a launch scheduled for spring 2013. This was a flight opportunity available to ASI, the Italian Space Agency, and as such it could only be offered to Luca or me. We'd discussed it often and openly, neither of us hiding the fact that we both wanted it unequivocally. Whoever was assigned would transition seamlessly from basic training to pre-mission prep and would reach the launchpad in less than three intense and challenging years. The other person, though, would languish in that state of pent-up frustration and anxiety typical of most astronauts who are waiting for their first spaceflight, and naturally uncertain whether the coveted opportunity will ever materialize.

It was no surprise when Luca was chosen. If I had been the highest-ranking officer at the time, it probably would have been me. This criterion was objective and indisputable, and it allowed ASI, at the suggestion of the Air Force, to make an otherwise difficult choice, since there were two of us apprentice astronauts, both of us equally prepared and chomping at the bit. Nevertheless, it was a great disappointment. Of course, there'd be a second ASI flight a couple of years later – that's why two Italians had been recruited into the Shenanigans in the first place. But you can't take anything for granted where space missions are concerned, and there were many unknowns. The only thing I could count on was that there would be years of agonizing uncertainty ahead.

None of this dampened my excitement about the prospect of two months' training at Zvezdnyj Gorodok. For everyone who lives and works there, it's just Zvezdnyj. Not all of the several thousand residents these days work in the field of space. Most of them walk each morning on the narrow road that leads from the main gate through the woods to the Tsiolkovsky stop, two simple platforms partially protected by shelters. From there, an *elektrichka*, the Spartan yet reliable electric train that connects Moscow to the smaller outlying towns, takes them to Yaroslavsky station in a little more than an hour. Then the extensive network of the Moscow metro gets them to their places of work and the universities. Some make the journey in reverse, and once they've entered Star City, they go through a second guarded gate to find themselves in the so-called technical zone. These are the specialists at the Yuri Gagarin cosmonaut training centre. Many of them are old enough to have known Yuri in person, while others are so young they don't even remember the Soviet Union. Because of the social and economic upheavals of the last twenty years, the generations in between aren't very well represented.

In Star City, it's not uncommon to run into someone you've read about in history books. If you hang around after working out in the cosmonauts' gym, maybe relaxing in the sauna, you may just discover that some of your more elderly companions flew on the first Soviet Space Stations in the 1970s, the Salyuts. In fact, many of the older cosmonauts still live here, in the apartments assigned to them in the Soviet era. And even if they've moved away, they still show up regularly. By chance, one day I ran into Vladimir Titov, well known for having experienced the only launchpad emergency escape ever, back in 1983. After Titov's rocket caught fire, thanks to the flight controllers' quick reactions and the impeccable escape system, his Soyuz capsule was separated from the launcher with an acceleration of 15 Gs, enough to save him and his crew. Another time, I passed an older lady who was walking to work in the technical zone. 'One of Tereshkova's back-ups for Vostok 6,' a Russian colleague explained matter-of-factly. Sometimes it seems like no one ever leaves Star City.

The main objective of those weeks of training was to become certified as 'users' of the Russian modules, which make up about one-third of the habitable volume on the Space Station. Users have the most basic qualification level: they must demonstrate generic familiarity with the on-board systems, know how to use the emergency equipment and be able to carry out everyday procedures, such as operating the electric switches and plugs, water dispensers and communication systems. Combining classroom theory, practical sessions in the mock-up and the incessant repetition typical of the Russian teaching philosophy, our instructors ensured that we became familiar with places none of us would visit for years to come. Even today, despite never having repeated these courses, I would still be able to find every one of the ten 10V sockets on board the Service Module with my eyes closed.

While most of our time was dedicated to the Russian segment of the Space Station, we mostly looked forward to sitting in the simulator for the Soyuz, the small spaceship that would one day take us to the International Space Station. I was thrilled to realize that we were studying how to operate the vehicle directly from the procedure checklists actually used by crew members, which you are not even allowed to take out of Russia. And I felt deeply emotional when I put on the Sokol: it was uncomfortable and too big for me, but I was instantly in love. I was looking out on the world for the first time through the helmet of a spacesuit.

The Soyuz simulator hall was large and full of light, with high ceilings and large windows. At regular intervals along its length four simulators protruded like lone mushrooms, each one a spherical orbital module sitting above a bell-shaped descent module. There were steps leading to an opening in the side of the descent module, and you could climb down through that onto the seats. This access does not exist in the real space vehicle, and the only way of getting into the descent module is through the orbital module. But it wouldn't be practical or safe to have to lower yourself from above during daily training in conditions of normal Earth gravity.

In the same simulator hall where we apprentice astronauts were just starting to get to know the Soyuz, a crew were taking their final exam, only a few weeks before their actual launch from the Baikonur cosmodrome in Kazakhstan. It was Yuri Petrovich who suggested that we sit in on the first part of the exam early in the morning, before we began our classes. We'd only met him a few days before, but it was already clear that he was our guardian angel. Equally esteemed by Russians, Americans and Europeans, who always respectfully called him by his first name and patronymic, he'd been managing the ESA office in Star City for a few years and had a reputation for solving every conceivable problem efficiently, and with meticulous attention to detail.

By eight in the morning a small crowd had gathered in the simulator hall. The three crew members, including NASA astronaut Scott Kelly, came in through a side room, where they'd put on their Sokol space suits. The suits are designed for the foetal position you assume in the Soyuz seats, definitely not for standing up, so the crew members were forced to walk awkwardly and bent forward. They proceeded towards the examining committee, half a dozen elegantly dressed officials lined up behind a small wooden table, on which lay sealed envelopes. The commander chose one of the envelopes, and all three of them signed it without having seen its contents. As Yuri Petrovich patiently explained to us in his usual gentle yet authoritative tone, the envelopes contained possible exam scenarios, each one describing a different sequence of malfunctions and emergency situations the crew would have to face. Responsibility would fall chiefly on the two cosmonauts, the commander and the flight engineer, respectively. Scott, who would be commander of the Space Station, would be carrying out a support role on the Soyuz and had received much less in-depth training than his Russian colleagues. Sitting in the right seat, he would not have direct access to the telemetry or the controls. In any case, Scott already had a fantastic career. As a naval aviator with more than 8,000 flight hours, he was about to embark on his third spaceflight after having served first as pilot and then as commander of the Space Shuttle.

When I went back to the simulator room late that afternoon following my classes, the exam was already in its final phase: in the simulation, Soyuz had undocked from the ISS, and the crew were preparing to re-enter the atmosphere. The simulator control room was chock-a-block; anyone who could possibly attend was there. Not a soul wanted to miss it. Instructors, teachers, cosmonauts and astronauts: all were watching, evaluating and discussing. I didn't yet have the knowledge I needed to really follow the exam, but going

44

by the instructors' proud comments, it seemed that the crew were doing very well.

Suddenly the measured tone of the conversation gave way to buzzing chatter and the commotion of people rising from their seats: the exam was over, the crew left the simulator to go and change and the committee retired to the adjoining room to consider the results. I was very curious and wanted to attend the debriefing, the discussion and analysis session that always follows a drill and today would include an evaluation of the astronauts' performance. I'd heard that the crew members are encouraged to discuss and to defend their decisions and actions with the committee in order to get a better score, and these discussions were rumoured to be very animated. But not this time. The crew, under the guidance of an expert commander, had acted more or less impeccably, and the few observations made by the committee were accepted without objection.

It was time to celebrate. Following tradition, the crew offered refreshments in the building next door, and we all – instructors, translators, doctors, cosmonauts and astronauts, friends and families of those celebrating – assembled around a groaning table. On behalf of the Shenanigans, I proposed a toast, the first of many that punctuated my time as an apprentice astronaut and would end years later, when a rocket sat waiting to launch me into space. Living in Star City, I'd quickly learned that making a toast is a deadly serious matter, and it's unforgivably careless not to be ready to propose one at the drop of a hat. Russian toasts run the gamut from witty anecdotes to sagacious, philosophical disquisitions, and consist of elaborate homages to courage, devotion to duty, friendship, love and family. It's not originality they value, but a heartfelt eloquence, so much so that, months later, you might still be complimented on an especially clever or moving toast. Above all else, a Russian toast is never short. Listeners wait patiently, in courteous silence and growing, but contained enthusiasm which finally explodes – after the traditional exhortation tri . . . . četyre (three . . . four . . .) – in a thundering 'Hurrah! Hurrah! Hurrah!'

I was to discover that final exams are only one of the many events that mark the passage of time in Star City, a place that breathes with the cycle and rhythm of the space missions. Following rituals observed for decades, the small local community gathers joyfully around the crew as they set off for space and reassembles to celebrate their return. When you come back, even after a period of some years away, it's easy to get back into sync, and there's a friendly greeting from familiar faces, as if you had been absent only for a short holiday.

Scott's crew had a week off after the final exam. Russians wisely allow this post-training break so that astronauts can recuperate and tie up matters pertaining to daily life on Earth before a long absence from the planet. The week over, we all got together again for a feast, this time first thing in the morning, the traditional goodbye breakfast held for the crew leaving for Baikonur. Time was short on this occasion. There were a few obligatory toasts before we were asked to quickly take our seats – despite the fact that there weren't enough chairs for everyone. Whether you sat on the windowsill or on your neighbour's knees, the important thing was for everyone to be seated for at least a few seconds before leaving, which was an essential part of wishing the departing crew a safe journey. No one touched their food. Barely twenty minutes later, with the crew on a bus headed for the airport, we Shenanigans, innocent novices that we were, returned to the breakfast room, intending to do the breakfast justice. But the table was completely bare. It seems this is the custom, and even now, years later, I still haven't solved the mystery of what happens to that feast.

Scott and his crew left for the Space Station a few weeks later, as planned, and I was surprised to find myself following the activities of the ISS particularly closely since, for the first time, one of the crew members was someone I knew. What I couldn't know then was that I would one day be welcoming Scott back to space for his final flight.

And Scott was the one to send me my first present from the Space Station a few weeks later: a lovely photo of Italy he'd just taken from the Cupola. It was on the occasion of our having completed our basic training. To celebrate, there was a simple ceremony in Cologne, upon our return from Russia, where we were presented with large diplomas certifying our official status as European Space Agency astronauts. It was a great moment for celebrating with our families, though not particularly emotional, because your feelings are much more intense when you've struggled hard to reach a goal, and I don't think anyone found the course that difficult, a testament to ESA's well-conceived astronaut selection. We might not have been the best – it's not reasonable to believe that such a long and complex process resulted in the perfect selection – but all things considered, it seemed we were good enough.

With our basic training at an end, we Shenanigans began to go our separate ways. Luca had already begun preparing for his mission, which would be named Volare, and the next ESA flight, which eventually became the Blue Dot mission, was intended for a German astronaut, so we knew Alex would be assigned soon. For Thomas, Tim, Andy and me, the prospects were much less certain.

I soon began to feel restless. An astronaut who hasn't yet been to space wants only one thing, and with every fibre of her being: assignment on her first mission. Though I knew my time would not come up for another couple of years at the earliest, I wanted to keep preparing if nothing else, training as much as I could as soon as I could. Around that time I learned that there was a way of beginning proper pre-mission training without a definite assignment, and that was to become ESA's reserve astronaut. The reserve astronaut would receive roughly the first year of the ISS training and would be able to replace a colleague from the same agency if they had to bow out of the flight in that first phase of preparation. It was an extremely remote possibility, but that mattered little to me. Above all, it represented the chance to train like an assigned astronaut, to qualify for spacewalks and for flying the robotic arm, to get to know about more of the Space Station beyond the European and Russian modules. And, last but not least, I would feel like I was genuinely part of the international astronaut community, with its centres of gravity in Houston and Star City. I volunteered without the least hesitation.

While I was waiting for the decision, I did a lot of public outreach. From Catania to Berlin, Milan to Paris, I travelled to over twenty European cities, mostly to speak in schools, though occasionally to participate in institutional events or TV programmes. As the months slowly passed, I felt increasingly impatient. Of course, outreach is an important duty for an astronaut, and it's normal for this to assume centre stage at times, but I found it difficult to cope with a situation in which I was talking about my past – my basic training – and my future – my flight on the Space Station – without knowing when my present would once again be interesting and challenging. Until at last one day, which was really no different from any other, there was a final signature, a final form, or maybe just a final agreement – and I was ESA's reserve astronaut.

I had no idea how steep and tortuous the road would be, or how long, but at least I was moving. I was now on my way towards a space mission.

6.

> At last I am settled in my function, no longer drifting into
> a faceless future . . . from now on my actions, one by one,
> will create the future.
>
> Antoine de Saint-Exupéry, *Flight to Arras*

Star City, 6 June 2011

I've never been afraid of getting old. The passage of time has always made me feel better about myself, and by nature I tend to
look ahead, curious about what will be, rather than look back
with nostalgia. It's true that I don't have the same boundless energy at thirty-four that I had as a girl, the inexhaustible energy
that makes you feel like you could take over the world – with
reserves left over to tackle a grizzly should you happen to meet
one. But every now and then I have days like this, when I suddenly feel the impetuous force of a river overflowing its banks. It's
my first day back in Star City.

Last night at the airport, I was filled with joy when I saw
Nikolai's kind face. He took me to the Profilaktorium, and
in the Russian way insisted on carrying my suitcases. The lady
on duty at the entrance gave me an enthusiastic welcome, almost reproaching me for having been away for so long. I didn't
even stay at the Profilaktorium the last time, but I did go there
almost every day to work online in the ESA office on the second floor. The office sits at the end of a hallway, behind a
glass door covered in stickers bearing the logos of the Soyuz
missions with European participation. Along the corridor

hang official photos of all the crews with Europeans on board. André Kuipers is on the wall with the 2004 Delta mission, but he is also physically present right now, as the Dutch labels on various food packages on the shelves of the small communal kitchen suggest. In a few months, André will leave on his second mission.

I settled into one of the four comfortable rooms that ESA rents permanently, to host astronauts and temporary staff. There's a bedroom with a double bed, and a day room with a sofa, a desk and a huge dresser in dark wood. Everything is a little old-fashioned, as in your grandparents' house. This morning, Yuri Petrovich knocked on the door and greeted me with a big hug and the keys to the lock for my bike. Anna, who looks after the ESA office along with Yuri, returned my pass to me with an affectionate smile, and took my passport in order to complete the necessary paperwork to register me with the authorities. I got my bike out of storage and stood for a few minutes at the door, taking in the clean air and the idyllic view of the lake, with its white gazebo and the narrow wooden pedestrian bridge that leads into the forest on the other shore. A few minutes' calm pedalling, and I reached the entrance to the technical zone. Star City was bright and warm, so different from the rainy grey of last autumn.

Even this morning's lessons in the Soyuz simulator are different. We're no longer sprinting ahead in some headlong, uninformed rush, just to get a taste of Soyuz operations, but we're now building competence slowly and patiently. I'm curious to learn everything, and I mean everything, about the Soyuz. The first step is mastering the names of things. With Ruslan, a young instructor, I go over all the components crowded into the descent and orbital modules, the two volumes of the small spaceship accessible to the crew.

At the end of the morning, with all those names swimming around in my head and only loosely attached to the things they represent – at least to the few initiated in the world who are familiar with the Soyuz – I say goodbye to Ruslan and ride my bike to the cafeteria to have lunch with Luca. I haven't seen him for months. As an assigned astronaut, his life is now ruled by what we call the trip template, an Excel sheet produced by schedulers from all the different space agencies and containing information on the movements of each crew member from the beginning of their training right up to the day of the launch. It resembles a colourful mosaic, with each square representing one week and various colours corresponding to training locations: light blue for the United States, yellow for Russia, green for Europe, purple for Japan and blue for Canada. Since there are only a few weeks of International Space Station training at EAC, Luca's not spending much time in Europe.

We meet up again that evening, along with the other astronauts who have a yellow square right now on their trip template. I meet Luca's crewmates: Chris Cassidy, whose mild manner hides his past as a Navy Seal and his being perhaps NASA's best spacewalker, and Karen Nyberg, a woman of many talents with a PhD in engineering and a passion for running. This time, instead of going to Cottage 3, we meet outside near a small building that has a working *banya*, or Russian sauna. There's a *shashlyk* outside, a traditional barbecue the Russians have adopted from the Caucasus. I speak little, simply enjoying the fact of being here in this place with these people. The pleasure I feel washes away what's left of my restive mood of the past few months – and I'm like a swimmer just emerged from the water who towels off and is now enjoying the sun and the breeze drying off any remaining wet patches of skin.

The evening over, I make the short walk back on my own. It's late, but at this latitude the summer days are long, and there's still a soft light. At the Profilaktorium, Yuri Petrovich has left a box in my room containing all the Soyuz manuals and checklists I'll be studying in the coming months. I pull them out and put them on a bookshelf, where they take up almost a metre of space.

It's time to get down to work at last.

—

Be careful what you wish for – it might come true. Or so the saying goes. My eight weeks in Star City that summer turned out to be interesting and fulfilling, but also extremely intense. My course was very compressed and designed so as not to conflict with the normal crew rotation. Within two weeks all the other astronauts had left, so there were no more suppers together in Cottage 3. I attended theoretical and practical lessons anywhere from six to eight hours a day and studied till late. Maybe I studied too much. I was pleased with the praise I received after my exams, but I was aware that not everything I was learning had a direct application on the actual operations of the Soyuz. It's well known that the Russians favour an extremely thorough theoretical preparation, which doesn't always correspond to the practical demands made on operators or the crew. But I didn't mind; on the contrary. My training had barely started, so I was full of energy, and the engineer in me was pleased to gain comprehensive knowledge of the systems, even about hidden aspects of the machine, ones I'd never have to use during a flight. In any case, it's always harder to learn something you've overlooked than it is to forget what you don't need. My in-depth studying would prove to be an ideal investment when I began the simulator sessions with my fellow crew members a couple of years later.

The exams were a very serious matter. For the more complex ones on the guidance, navigation and control system, the classroom was always teeming with instructors who seemed to be competing with one another to ask the most difficult and convoluted questions. I'd answer from the teacher's desk in a scene almost embarrassingly reminiscent of school, and after a period that could vary from half an hour to two hours, depending on the complexity of the subject, I'd be asked to wait for a few minutes in the corridor while my score – on a scale of one to five – was decided.

There was a lot to study, and the only distraction I allowed my-self was a daily run along the paths that criss-crossed the forest and around the small lakes in the vicinity. As I wrote to a friend at the time: 'I feel like I'm on a spiritual retreat: just me, the manuals and the birch trees.'

These were peaceful days, full of satisfaction for someone who was building her dream, brick by brick.

7.

Star City, 18 July 2011

The privilege of finding myself in the commander's seat is easily explained: I'm the only one here. I'm sitting in the Soyuz simulator; the seats to my right and left are empty. The one in the middle is the most comfortable – or should I say, it's somewhat less uncomfortable than the others, since it's slightly bigger. I smile to think that the Russians use the word *kreslo*, literally 'armchair', a word that suggests an entirely different level of comfort. In any case, these seats don't have to be comfortable. They're meant to protect the crew as they land, at the moment of impact with the ground. That's why they have such a peculiar shape: the outside is a metal shell mounted on a shock absorber, and the inside is a blue seat liner, custom-shaped to the back and head of each occupant. It's not even appropriate to call them 'seats', since you don't so much sit in them as lie down in them – in a foetal position, that is, with your knees bent against your chest and your feet resting on a small support just a little beyond your buttocks. Never before had I been so happy that I'm not very tall.

You have to wear the Sokol spacesuit in these seats, but you don't always wear it during training, and certainly not in this initial phase. When we train in plain clothes, we lay a thin mat over the seat, both to be more comfortable and so that we are not too far from the control panel. Before taking my place, I grabbed one of these mats, as well as an extendable baton, a primitive but indispensable tool for reaching the controls from the central seat. That's how we fly in the Soyuz, using the baton to push large

54

buttons separated by wire cages to stop you sliding from one command button to the next by mistake. Of course, in the age of touchscreens and popular voice-controlled personal assistants, it's rather endearing to find yourself controlling a spacecraft with a baton, like a teacher in front of an old-fashioned blackboard. But I really love it. I'll admit, I even take a bit of smug pride in it, like when I was a child on the ski slopes. I practically grew up on them, and I always watched tourists on their ski trips with amused superiority. They were all kitted out to the nth degree but couldn't ski if their lives depended on it. I might not have had all their kit, but I felt like I owned the mountain. Today, in the same way, I leave the stylish headphones to others and happily put on the *shlemofon*, the black cloth headgear with integrated earphones and microphone. The astronauts are divided between those who are convinced they still bear traces of Yuri Gagarin's sweat and those who sensibly realize that there's no way we'll ever know for sure.

So here I am, sitting in the commander's seat. I will definitely never be commander of the Soyuz – that role is reserved for Russian cosmonauts – but during these weeks I'm training as flight engineer, and in this initial phase, the preparation is identical to a commander's training: there are around 500 hours of lessons in both theory and practice, organized in courses dedicated to each system, all of them followed by a rigorous exam.

I'm in the simulator for one of the last practical lessons on the guidance, navigation and control system. It's been a tough course, but one of the most interesting because it's allowed me to understand how the Soyuz – with its combination of sensors, engines, software, and data sent from Mission Control – is able to take astronauts to the Space Station and bring them safely back to Earth. Because – let's be clear – the Soyuz flies on its own; if all goes to plan, it doesn't need a pilot. In case you don't believe this, consider how the twin cargo vehicle, the Progress, takes supplies to the ISS with no crew on board. Almost everything I'm learning to do has two objectives: I need to maintain a constant check to ensure that all systems are functioning well on automatic, and I need to know how to intervene manually in case they aren't. The crew's presence increases system redundancy, since it introduces the possibility of manual control. At the same time, it's because of the humans on board that such redundancy is particularly important, as the crew is considered more valuable than cargo: an obvious fact, maybe, but one for which astronauts are always very grateful.

Learning how to react quickly and efficiently to system failures and emergency situations is reserved for advanced training, and that will follow when – hopefully – I am assigned one day to a mission as the flight engineer. For now, I need to learn how to understand the language of the Soyuz. It speaks constantly to me, giving me countless details, moment by moment, of the simulated flight and how the systems are doing – temperatures, concentrations, speeds, angles, pressures . . . Dozens of vital parameters illustrating the vehicle's state of health and the status of the manoeuvre in progress. I just have to know how to read it. I have to admit that the human/machine interface on the Soyuz isn't very intuitive, no doubt about it. The screens are for the most part populated by strings of letters and numbers which meant nothing to me until a few weeks ago. But now they speak to me in a precise and comprehensible language. Now, for example, they're telling me in crystal clear terms that the Soyuz is spinning in order to find the Earth's horizon, using its infrared sensor. Once it finds it, it will keep turning to orientate its 'belly', from which the periscope protrudes, along the local vertical, the straight line that connects our position in orbit with the centre of the Earth. If you were at that line on the Earth's surface, you would see us flying directly overhead.

As the Soyuz spins, I can see the Earth slowly entering the periscope's field of view, like a planet that appears before your eyes as you slowly sweep the sky with a telescope. At the same time, I watch the angles on the infrared sensors carefully, as I've been taught. The first sensor has already failed, and I suspect that my instructor will disable the reserve one as well. Just as I expected, he does so slyly, without setting off any alarms. There's nothing for me to do but take manual control in order to complete our orientation.

The manual controls on the Soyuz are fairly minimal: two knobs the size of an apricot, positioned on each side of the central viewer and mounted on an extendable track. The one on the left controls the translation and is useful if you have to manually dock to the Space Station. Right now, however, I use the one on the right, which controls the rotations. From its resting, retracted position, more or less at foot level, I pull it towards me, positioning it so that I can grab it easily, even as I lie on my seat.

My task is pretty simple, to tell the truth. I deflect the command knob until the circle of the Earth's horizon is perfectly centred in the viewer, which shows that the periscope is pointing towards the Earth along the local vertical as desired. However, the surface of the planet runs across the viewer in a roughly right to left motion: if we were a car, we might say that we were sliding sideways along the road. So I rotate the manual control anticlockwise, as if it were a doorknob, until the continents and oceans alternating in the periscope's field of view run across the viewer from the top to the bottom. We've straightened out on the road. I'm satisfied to hear the instructor say that the residual error is within acceptable limits. You have to train your eye to be very precise, since small errors can have serious consequences. On a spaceflight it's crucially important that the engines are turned on while the vehicle is in the correct attitude, or you may end up on an unexpected orbit – or, if you're returning to Earth, land on the wrong continent. A large part of training on the Soyuz could be summed up like this: turn the engine on at the right time and in the right attitude, whatever the circumstances.

The next task I'm practising involves orientating the solar panels on the Soyuz towards the Sun. I'm once more at the manual controls, this time looking through the periscope for our star. The Sun appears in the viewer, which is conveniently covered with a protective filter. I perform a series of rotations and watch the voltage values in the solar panels. They're increasing: more than 30V, and the mission is accomplished. It's now necessary to ensure that the Soyuz starts spinning so that in its orbit around the Earth it will keep the solar panels facing the Sun, without the need to actively control the attitude.

This simple and effective method, called 'gyroscope stabilization', is commonly used for satellites, rockets, missiles and projectiles. When there is a crew on board, however, it also causes some inconvenience: it's easy to imagine how unwelcome continual rotation is to an astronaut's vestibular system, especially when they've just arrived in orbit and are often afflicted with space sickness. I was told that it's a good idea to position your sleeping bag with your head as close as possible to the axis of rotation, so it's more or less stable, and you can alleviate the symptoms of nausea.

Of course, here in the simulator, nothing is really turning apart from the fan for air circulation. Enough for today. My instructor is satisfied and pronounces the lesson finished, slightly ahead of time. When I've extracted my legs, I'll have time to join him and the two simulator technicians for coffee and biscuits before I run to my next class.

———

At the cosmonaut training centre, you didn't go out for a coffee because there weren't any cafés. You didn't go to the machine, either, since the only one available was in a private lounge reserved for astronauts and mostly deserted. It was instead established tradition to have a kettle and some instant coffee in every office or workspace, and customary Russian hospitality meant that you were frequently invited for a chat and a steaming mug of coffee. I'd drink it happily, and I didn't even miss espresso or American filter coffee. Habits and tastes are, of course, flexible and have a lot to do with context. In any case, I'd be drinking instant coffee for months on the International Space Station, and these days when I recall filling a sachet of instant coffee at the water dispenser, taking it with me to my first afternoon activity and sipping it while I read the procedure, I'm instantly transported back to the daily routine on the Space Station. I even drank a few espressos in space, something I would have found quite incredible back then.

My routine that summer – lessons, study and runs in the woods – was temporarily disrupted by the arrival of a crowd of new faces. They'd come for the traditional welcome home ceremony for the crew of Expedition 26/27, the most recent mission to have returned from space. After they landed in Kazakhstan a few weeks earlier, crew members Paolo Nespoli and Cady Coleman had boarded a NASA plane headed for Houston, while Commander Dmitri Kondratyev had been taken back to Moscow. All three of them had now pitched up to receive the Star City welcome. I had time to see the beginning of the festivities around Yuri Gagarin's statue, where a small crowd of staff from the training centre had gathered, along with locals, among them several pupils. In a ritual I, too, would follow to the letter when I returned from space, Cady, Paolo and Dmitri placed flowers at the feet of Gagarin's statue and then followed the group in a brief parade along the wide road leading to the city's theatre. There they listened for hours to the warm words spoken and accepted gifts, including an amazing number of flowers.

I was especially happy to see Cady again, a veteran of two previous Shuttle missions with a PhD in chemistry and a passion for the flute, known for having played in the first ever Space–Earth musical duet. Cady is more sociable than any other astronaut I know – in fact extremely so. She makes friends very easily and provides generous and practical help and advice. She is particularly keen to help other women, so much so that, as well as offering her friendship from the start, she invited me to stay in her house in Houston when she heard that I'd be going to the Johnson Space Center at the end of September. For some time, in fact, a new line for reserve astronauts had appeared on the trip template, that compact representation of dozens of lives, interwoven and in constant movement. The line wasn't running towards any clear goal – there was no launch date – but it was nevertheless cause for joy.

At last! There I was, on that mosaic of coloured squares, hidden behind the initials CT.

8.

Houston, 27 September 2011

NASA's Neutral Buoyancy Laboratory – everyone calls it the NBL – is only about 500 metres as the crow flies from where I'm staying at Cady's house, so close that I can easily make out the flag flying above the treetops. But to get there takes a little longer, a few minutes by car through a labyrinth of residential streets lined by one- and two-storey family houses with neat green lawns and a couple of trees in their front gardens. I even see a few people walking and running, parents taking their children to school.

My first day of training starts here at the NBL, an imposing white building a few kilometres north of the Johnson Space Center. I well remember the awe I felt last year when I visited with the other Shenanigans. Even before we heard about all the complex operations that take place here, I was struck by the remarkable size of the facility. The heart of the NBL is a giant pool large enough to submerge an entire four-storey building and containing as much water as ten Olympic pools. At the bottom are the non-Russian modules of the Space Station, faithfully recreated to scale.

I'm looking forward to plunging into that underwater world, but it's not yet time. I won't even see it today. At the building's entrance, I ask for the suit room, and I'm directed down a long lateral corridor that runs along the imposing pool walls. As I walk the corridor, it's like being in a basement because the pool deck, and all its frenetic activity, is several floors up. At the end of the corridor I ring a bell and a young woman comes to open the door.

'Hi, I'm Robin, your suit engineer today.' Of course she means the EVA suit, which is used both in space and for underwater training. Its official name is Extravehicular Mobility Unit, or EMU, but the community calls it simply 'the suit'. It's been used for thirty years without significant changes, and it doesn't look very different to the suits used by the Apollo astronauts on the Moon. Few objects are as emblematic of the space programme as this white shell, which protects the fragile human body in the hostile environment of open space. Every astronaut dreams of being able to put it on one day, up there in orbit, and I am no exception.

Robin escorts me through a room halfway between a lab and a tailor's shop. I slow down, watching a group of technicians with voracious curiosity as they busy themselves with torsos, arm-pieces and leg-pieces, boots, gloves and helmets. The NBL suits are regularly dismembered here for maintenance, or to re-size them for someone else. In fact, no astronaut has a made-to-measure suit. Before every training session, the technicians assemble the suits piece by piece, taking into account each astronaut's individual measurements. What size are the single components, such as the torso or the forearm? How many sizing rings are inserted between a leg and a boot? What's the configuration of the numerous adjustment points that lengthen or shorten parts; to ensure, say, that the elbow and knee correspond more or less to those of the human inside the spacesuit? Today, we'll determine the answers to these questions.

I leave Robin after a couple of hours and more than eighty measurements – twenty-six for my gloves alone – and hurry to get to my next class, which is marked on the agenda as 'small EVA tools'. The young instructor introduces himself as Daren, and then theatrically gestures towards a table large enough to seat twenty. It's entirely covered with objects, some of whose functions are more or less obvious, and others more mysterious: I'm guessing they're the small tools. Suddenly, the three hours at our disposal seem like hardly any time at all. The sheer quantity of kit spread over the table and chairs is intimidating. But I'm really curious: here are the tools and aids with which the Space

Station was assembled in orbit. Or at least some of the tools, since I note there's a second lesson on the agenda scheduled later in the week: 'large EVA tools'.

To tell the truth, today's tools aren't even that small. They actually can't be, since they have to be held through the stiff, thick gloves of the EMU suit, which substantially reduce your sensitivity, fine motor skills and even the strength you can exert with your hand. Daren explains how each tool works, and I in turn try handling them, taking pictures, furiously writing down notes. I'm fascinated by the way in which the world of EVA has given rise to a universe of tools that have no equal in any other workshop in the world. Spanners, screwdrivers, torque multipliers – these are relatively common tools, but each one, when designed for use in outer space while wearing an EMU suit, is a thing of genius.

Near the end of the class, Daren suggests an exercise that he assures me will prove very useful when I have my first run in the pool in three months' time. Together we configure the so-called mini-workstation, a metal plate with a T-bar that's firmly secured to the front of the spacesuit. Astronauts put hooks on it and the tools they use most often, in line with the requirements for their EVA and personal preference. Some put as much as they can there so they don't have to scrabble around in the toolbag so often; others do the opposite, limiting themselves to the essentials in order to reduce clutter and the risk of getting caught on something. There are two tools, however, that are so useful and so frequently used that they are almost always carried on the spacesuit: PGT and BRT.

The PGT, or Pistol Grip Tool, looks like a laser pistol from a science-fiction film. It's the outer-space version of a battery-operated screwdriver and the main tool with which crews have assembled the Space Station over the years, bolt by bolt. It's usually carried on the right side and attached to a swing arm in such a manner that, when not in use, it can be rotated out of the way in a transport position close to the suit. As with all objects taken into outer space, it's attached with a retractable tether. This is the golden rule with EVA, Daren explains: everything must always be secured in some way to the astronaut or to the ISS structure to stop it from floating away, which would possibly mean losing an important tool or piece of equipment. Moreover, in keeping with the laws of orbital mechanics, objects that appear to have gone for ever sometimes have the bad manners to turn up again during the next orbit, and risk hitting the Space Station. And it would be particularly unfortunate if the lost objects were humans, which is why astronauts always have to be attached to the Station in two different ways: the first is the safety tether, wound in a coil that slowly unwinds as you get farther from the anchor point; and the second depends on the situation. If you are moving from one place to another on the ISS, it can be simply your hand gripping a handrail. When you reach your place of work, however, it's usually necessary to have both hands free. You can simply put down a local tether, so you won't float away, but this won't stabilize your position or prevent you from slowly spinning out of control when you let go.

A better option is the BRT, or Body Restraint Tether, a heavy tool that looks like a thick snake and is carried folded at your side when not in use. The free end is a metal clamp which looks like the snake's menacing jaws when it's open. Daren cautiously places a finger in it from the side and presses the base of the jaws, which abruptly shut with a metallic sound. That's how the BRT is fixed to the standard handrails on the ISS. By rotating a ring, it's then possible to tighten the body of the BRT, which is made up of a series of metal spheres on a cable; when pulled against each other with enough tension, they cannot move: for the time necessary, the snake is petrified and thus the astronaut's position is also fixed, and they can remain at the right distance from the bolts, electrical or hydraulic connections, or whatever else they might need to work on with both hands.

Houston, 30 September 2011

It's a terrifying scene: I'm outside, in EVA, and I've become physically detached from the ISS. Perhaps I lost my grip on the handrail, or failed to properly secure a hook. If this were to happen, we have the safety tether, which should have saved me and hauled me towards the anchor point, but something has clearly gone wrong, and now I'm drifting away, tumbling. My only hope of turning back towards the ISS now is the SAFER, the integrated jetpack on the back of my EMU suit. With its twenty-four small compressed nitrogen thrusters arranged in four clusters at the level of my shoulders and my hips, the SAFER guarantees a small reserve thrust with which I can rotate and turn back towards the Space Station.

If I were really in space, I'd have to try to stay calm now and focus on my first priority: lowering the T-bar on my mini-workstation and extracting the SAFER hand control module. I'd have to turn the swing arm at my right side to reach the extraction lever. With that action, the spring mechanism would extract the hand control module from the compartment on my back and bring it round to my field of view. To be honest, I don't think you can assume that you could do all this in a timely manner, especially in a situation of imminent danger, but I'll have a chance to think about this in the future. Today I'm already holding a joystick with three toggle switches on it. I'm about to try my hand at the SAFER for the first time in the Virtual Reality Lab at the Johnson Space Center.

I'm immersed in a simulated environment that reproduces the outside of the Space Station in every detail. My head, chest and hand movements are tracked by dedicated sensors and the image projected on the viewer in front of me is modified accordingly. At the beginning of the simulation, I was positioned outside the Airlock, the small module through which one emerges into outer space. But I was immediately projected away from it. The ISS now appears in my field of view every three or four seconds, and each time it's further away. I try to orientate myself, to understand the direction I've been thrown in, but I am not yet very familiar with the outside of the Space Station: I see bits of the solar panels, segments of the truss, a docked Soyuz or Progress – it's impossible to tell which without detecting a periscope, the only obvious distinguishing feature between the two. After exactly thirty seconds, which simulate the time considered necessary to take out the hand control module, I'm given permission to begin the self-rescue procedure.

The first step is to stop the uncontrolled spinning, and for this, the SAFER makes my life easy: with one simple command I can automatically bring all rotation rates to zero. Now I have to find the Space Station: I turn slowly until it appears in my field of view and I try to align myself with the point from which I became detached, the Airlock. If I can manage a perfect alignment, a forward impulse would allow me to cancel my velocity of separation from the ISS in the most efficient way. It's important to be precise, since there's not much extra propellant. The nitrogen supply will forgive a little imprecision when judging the trajectory, but not larger errors: too many wrong or unnecessary thruster firings, and you may find yourself with an empty tank and no way of getting back to the ISS. There are no second chances.

At this point I'm a few metres from the Airlock and I estimate that I'll soon be able to grab one of the handrails. I look down, and the virtual control module indicates that I have about 20 per cent of the propellant remaining. Not bad for a first try. Of course, I know very well that the starting conditions were fairly simple. Before I leave some day for the real ISS, I will have to demonstrate that I know how to manage much higher rotation rates and separation velocities. As always, it's one step at a time.

Houston, 3 October 2011

I let all the air out of my buoyancy control device and sit on the bottom of the pool, close enough to see what's going on in the Airlock, but far enough away not to hinder the team of support divers. Over and around me is a world of water and metal: no matter where I look, my gaze meets the metal grilles of the pillars that support the imposing replica of the Space Station. Just a few metres from the bottom are the white cylinders that simulate the pressurized modules; a little above is the truss, its top almost breaching the surface of the water. Plastic tubes stand in for electric cables and hydraulic pipes; segments of yellow railing outline the translation paths along this submerged Space Station. In space you have the same segments in the same positions, except that they are sometimes dented from impact with a micrometeorite.

I'm used to silence during underwater dives, but here, loudspeakers relay communication between the astronauts in their suits and the instructor, who follows the activities from the control room with the aid of a live video feed. For each suit there's a camera installed on the helmet, and a second one is held by a diver. There are about ten divers in all, two assigned to each astronaut in order to ensure their safety and to help them move around as necessary. The others are there to support the simulation, bring tools and configure parts of the mock-up in real time.

From my standpoint at the bottom of the pool, I watch the astronauts move around deliberately, as in a slow-motion film. Their every movement has to be negotiated, taking into account the rigidity of the suit and gloves and their restricted field of view, not to mention the fatigue accumulated after almost six hours underwater. But the measured pace is also strategic. 'Slow is fast' is EVA's mantra: if you work unhurriedly, you'll get there faster in the end.

I've asked to scuba-dive in the NBL as often as possible so I'll be well prepared for my first run in the suit next spring. I want to become familiar with the ISS mock-up, I want to practise typical translation paths again and again, I want to know how to find the handrails with my eyes closed. In all honesty, it may be that I just want to spend as much time as possible in this underwater world, which seems the place on Earth that most resembles space.

Houston, 4 October 2011

Suspended as it is over the pool deck, the NBL control room offers a fascinating show. You can admire the huge pool through a large glass wall, and when the water is undisturbed, the metal world submerged in it is clearly visible. There aren't any suited astronauts in the pool today, but there is some activity there, with several pairs of divers getting ready and periodic service announcements coming over the loudspeakers. Sarah, one of NBL's most experienced divers, explains to me that her colleagues are configuring the mock-up for the next training sessions. As she points out, the NBL models aren't wholly faithful to the originals. After all, there isn't much need for a working pump in the water, for example. The astronauts will never be asked to repair a broken one, only to replace it with a spare that's stored outside the ISS. The model must faithfully reproduce the layout and the interfaces astronauts have to work with, such as handles, bolts, electrical and hydraulic connections.

It's Sarah who takes many of today's briefings on training in the suit. She begins by telling me what a typical day is like at the NBL. Astronauts arrange their tools on the pool deck, before a medical examination at 7.30, followed by a briefing with the entire team. Then they suit up. Around 9.00 they go underwater, and before anything else, the two divers assigned to each astronaut perform the weigh-out at the bottom of the pool. Sarah goes over this carefully: a proper weigh-out is essential to allow for the most efficient training and to avoid excessive fatigue. The more accurate it is, the more realistic the simulation of weightlessness. The divers must first establish neutral buoyancy, adding or removing weights and small foam blocks from the suit to neutralize any tendency on its part to sink or rise spontaneously. At the end, the weight should be perfectly balanced by the buoyancy. Up to this point, things are relatively easy. Yet as Sarah explains, it's at the next stage that the weigh-out becomes an art. In order to simulate being in space and the ability to work in all positions and orientations, it would be useful to have a neutral attitude – the suit, with the astronaut inside, should be able to turn in any direction without meeting resistance, and to maintain any position without rotating spontaneously. Though that is, unfortunately, an unattainable ideal, the divers do everything they can. The weights and foam blocks are placed in special pockets on the chest, back and ankles of the suit in order to weigh down or lighten parts as necessary.

Sarah does not conceal the fact that the weigh-out is easier and more effective for bigger astronauts who fill out their EMUs. They have a smaller air bubble inside their suits and are kept in a stable position simply by the bulk of their own bodies. Because, let's face it, even if the suit is in neutral buoyancy, the astronaut isn't floating inside it. If you go upside down, for example, you feel your entire body weight pressing down on your shoulders, and for this reason astronauts are prohibited from working in that position for more than ten minutes. Despite that, shoulder injuries are not, alas, uncommon, because of the unnatural position the joints are forced to work in. Tom, a doctor and an astronaut colleague, takes over from Sarah to illustrate a few recommendations. The first thing to ensure is proper weigh-out, because the closer you get to neutral buoyancy and attitude, the less effort you need to make to maintain the desired position. It's also essential to strengthen the stabilizing muscles of the joints, and above all, the small rotator cuffs: I can surely ask the trainers in the Astrogym to help me with that. Ultimately, it's very important to achieve the best possible suit fit. Working in an ill-fitting suit is like taking a long walk in unsuitable shoes that cause pain and blisters. This image gives the right idea – but I suspect that getting a good suit fit is much more complicated than buying the right pair of shoes.

Houston, 5 October 2011

The first step in achieving a good suit fit is to choose your gloves. Even that proves harder than you might think. This morning, in fact, I've come back to the room where Robin took my measurements last week, and it's official: I have strange hands. Steven, my suit engineer for the day, hands me a chart showing all the available gloves as dots, arranged according to size. My hand measurements are also condensed in a dot: a lonely dot in an empty area of the chart. As it happens, there aren't any gloves that correspond to my hand size. I would probably be a good candidate for a pair of custom-made gloves, and that would help to fill that blank in the chart. But they don't make that many, and there's already a long list of astronauts who've been waiting for some time. Since I'm the last to arrive and haven't even been assigned to a mission, I'm very low priority.

So we have to choose not one, but two pairs from what's available: one as my primary pair and a second back-up pair. I feel a bit apprehensive because I know it's a tricky decision. In space, as underwater, you're not walking, so your hands are always in action, whether they're moving hooks and tools, seizing handholds and handrails, helping you move or change direction. Gloves must ensure sufficient sensitivity and dexterity, and they should not lead to unnecessary effort, nor cause pain and friction. Fatigue actually makes itself felt in the hands more than anywhere else. Someone told me that a pool run was like spending six hours trying to squash a tennis ball in your hand. They may have been exaggerating, but probably not all that much. Forewarned is forearmed. I won't have my first run in the NBL until next March, but for several months now I've been carrying one of those stiff rubber rings with me everywhere I go, squeezing it to train my hands.

The five gloves Steven singled out as promising candidates are neatly spread over the table. The technicians pick up the first pair they've prepared for the trial, attaching each glove to a suit forearm, and then they help me to put it on. Steven asks questions from his long checklist. Do I feel good contact with the gloves both at the fingertips and between the fingers? Can I bend my thumb easily or is the joint in the wrong position? Is the metal bar that stiffens the gloves in the right place on my palm?

After some small adjustments, we proceed to trying them on in real conditions, replicating the stiffness of the gloves when they're used in space at 4.3 psi of overpressure compared to the surrounding vacuum. We could contrive a way of inflating them to achieve this overpressure, but it's easier in this case to lower the surrounding pressure. The result is the same. Let's say you wanted to inflate two balloons with helium until they popped. You could fill one with helium from a pressurized canister until you got the desired effect, and you could put a small amount of helium in the other, tie a knot and let it go. As it rose and the atmospheric pressure slowly decreased, it would inflate until it popped. If we disregard manufacturing differences and variations in temperature, the balloons would burst when they reached the same overpressure with respect to their surroundings. The absolute pressure is not important. So in some cases, such as in the pool, we inflate the suit until it has 4.3 psi more pressure than the surrounding water; in other cases, like now, we lower the surrounding pressure by 4.3 psi. The gloves swell and grow stiffer in the same way.

Of course, we can't lower the pressure in the whole room, as that would be complicated. Instead, we use a glovebox, a sort of cylindrical aquarium without water, and with two circular openings spaced to allow you to insert your hands comfortably. Wearing my EMU forearms with the gloves, I insert my hands until the metal rings at the end of each arm are securely wedged against the flanges of the glovebox openings. As soon as the technicians turn on the pump, the internal pressure begins to fall, and the difference in pressure locks the metal rings in position, creating a hermetic seal.

I can feel the gloves beginning to inflate. I sense them less and less as a sheath squeezing my fingers and more as a shell with rigid walls in which I have plenty of space to move my hand. Too much space, I suspect. When the pump stops, I try making a fist and manipulating some of the hooks in the glovebox. The strength I need for this is remarkable: I was warned, but it is much more difficult than I expected. I have a hard time imagining how I could work like this for six hours in the pool. I try to stay optimistic anyway, reminding myself that these are the first gloves I've tried on and maybe others will fit my hands better and be less tiring to use.

After more than two hours of work, the situation is not encouraging: none of these gloves seems to be an ideal solution. All the same, we select the least problematic pair for me to wear, inserting thick liners and a soft foam cushion on the back of my hand in an effort to fill up the empty space. It's a temporary fix. Steven will look into other options, and we'll try them out in another session.

Being the cause of extra work makes me feel uncomfortable, all the more so because I've only just arrived in Houston and I really don't want to gain the reputation for being complicated. Happily, Steven shows no sign of irritation and actually apologizes for the difficulty. And then there is Sandy: pragmatic, as always, she encourages me not to settle for less, and to try out as many gloves as I need to.

I only met Sandy a few days ago, but I've known about her for years. She has a PhD in material science and a passion for soccer, so much so that she played on her college team. A veteran of three space missions, one of them a four-month expedition on the ISS, only a few months ago she was on the crew of STS-135, the last Space Shuttle flight. When she heard that I would shortly be starting my EVA training, Sandy announced without the least hesitation that she would come with me for the glove fit check. If you don't have any experience, she explained, it's difficult to know how to ask the right questions. She wasn't speaking like someone doing me a favour, but as someone simply stating the obvious. Like Cady, who welcomed me into her own home, Sandy is another NASA astronaut spontaneously offering me her friendship and generosity. I can't be sure that I'll be going into space one day, but one thing is certain: simply being a part of this community is a privilege.

Among the gloves I've tried on today there's also a pair marked MG, for Magnus: years ago, they were custom-made for Sandy. They're the only women's gloves that were offered to me and they'd be perfect if not for the slender fingers. I think that's the problem: my fingers aren't tapered like most women's, and yet the men's gloves are too roomy. And to think that I've always loved my 'craftsman's hands', as one of my schoolteachers called them. I've never worn rings or grown my nails, since I always liked the way my hands conveyed strength and practicality, a sort of external sign of my desire to mould my own life. I never dreamed that they'd one day end up as a lonely dot on a NASA chart.

Houston, 11 October 2011

This is how an overturned turtle must feel. Lying flat on the grey carpet in a spacesuit weighing more than 100 kg, with my arms extended towards the ceiling, I very much doubt whether I can stand up without help. Fortunately, I don't have to try. In a few minutes, the two technicians who gently let me down will help me to get back up.

I'm wearing a Snoopy Cap, a cloth hat with earphones and a microphone in it. It takes its name from the cartoon dog because of its characteristic colour pattern, which became famous at the time of the Apollo missions: black over the ears, white central strip. I can hear Steven's directions through the headphones; he's standing beside me, asking questions. I wouldn't be able to hear him without the radio system, since I'm wearing a helmet, and the suit is completely sealed. After a second fit check for the gloves – we chose a new pair – today we have to select the different bits which the technicians will use to build my suit before every pool run. 'Build the suit' is precisely what they say. Steven has prepared an initial configuration based on the measurements we've taken, and with an overpressure of 4.3 psi, we're evaluating small modifications in order to optimize the work envelope, the volume in which I can work effectively with my hands. A few minutes ago, it was time to lie down on the ground.

So I'm lying on my back as described, with my arms stretching towards the ceiling, and Steven is asking me whether my hands are still in the gloves. They're not. I don't fill out the suit's torso – there's a lot of empty space in there. When I was standing up, I could move forward and still ensure that my hands stayed firmly inside the gloves, with my fingers right down at the ends. But when I'm lying down on the ground, I fall towards the back. Steven takes note. The solution is not to shorten the suit's arms – I need the reach – but to add padding to fill up the free space behind me and stop me from falling backwards. You do float inside your suit in space, but not much in the NBL, where I've got many hours of training ahead of me, and some of them will surely see me lying on my back. At least there's no doubt about the size of the torso. It's a rigid component that comes in three sizes: medium, large and extra large. Medium is perfect for colleagues such as Luca and Alex, who are both taller than I am and have wider shoulders. It will have to be OK for me, too.

When he's finished asking questions, Steven gives the green light to the technicians, and they help me to get up. They support me while I awkwardly go back to the donning stand, a structure with designated points of attachment which are used on Earth and in orbit to keep the EMU in a fixed position so you can get in and out of the torso piece. It's time to take off the suit and let Steven and the technicians try out some padding and adjust the length of arms and legs.

First of all, the overpressure must be slowly decreased. Then, to get complete equalization with the ambient pressure, one of the technicians removes a glove. He's asked me to make a soft, continuous sound to ensure that my airways remain open, preventing damage to my eardrums. When they've removed my gloves and my helmet, the technicians move on to detach the lower part of the suit, pelvis and legs, which is connected to the torso by a metal ring at the hip.

The last thing I have to do is wiggle out of the torso. The technicians hold the suit's arms up in the air, and I crouch down, taking my arms out one at a time. 'Too easy!' jokes Steve, the veteran astronaut who accompanied me today. I actually did get out of the suit quickly and easily, but it's a difficult procedure for many astronauts. At least in this respect, having to use an over-sized suit is an advantage.

—

I am impatient by nature, and I have a tendency to want to finish things quickly, though not in this case. The fit check went on until mid-afternoon, several hours later than planned. The following week I had an opportunity to try the suit again in my first Prep and Post class, a simulation of the long, complex operations in the hours before and after going out into space. It was a chance to familiarize myself with the suit-up procedure and how to move inside the EMU. 'Don't fight the suit' is a phrase frequently heard when you start training in the pool. Don't fight the suit, because the suit always wins. So you learn to play along with its limitations and make the most of the little freedom of movement that its joints permit.

At the end of that Prep and Post day, I allowed myself a glass of wine at Chelsea, one of the NASA community's favourite places for an after-work drink. The sweltering, oppressive Texas summer, which you can only escape in air-conditioned places, was finally over, and the mild October temperatures were ideal for sitting out on the terrace in a t-shirt, enjoying a view of the bay and the soft, late afternoon light.

I'd invited a few friends and acquaintances for a goodbye drink. Tomorrow was the last day of my first training trip at the Johnson Space Center. Although there isn't such a strong community life in Houston as there is in Star City, I had begun to lay the foundations of relationships that would prop up my life in Texas over the next three years. I'd enjoyed a lot more free time there than I would during my numerous return training trips. This first month hadn't been very demanding, since no one was expecting anything of me or my performance just yet. Still, I was very conscious of the fact that you start building your reputation right away, and a reputation is no less important than exams or formal evaluations. I knew that each one of the instructors I'd worked with – several dozen by now – had gone back to the office with an opinion of me: good or bad, they'd have been sure to share it with colleagues. Everyone is curious about new astronauts – other astronauts as much as or more than anyone else.

I left Houston hoping I'd made a good impression, or at least no significant gaffes, and returned to Star City to continue my training as a flight engineer. I took with me a new awareness of the surprising level of technical and operational complexity behind every spacewalk, something I could only imagine at that stage of the game. I'd put my toe in the water, but I had no way of gauging the depth of the sea.

9.

Star City, 18 January 2012

Thomas is the first to get into the battered old descent module lying on its side in the packed snow. I get in after him to find him huddled in a bizarre position above the control panel. This is the first time we've worked together since completing our basic training, and I'd forgotten how efficient he can be. I squat in a corner, hoping to make enough space for our commander, Sergey, to come in and close the hatch behind him. A quick radio call – and the drill begins.

We've been told more than once that the best way to avoid hypothermia is to stay dry, which means we need to move slowly so that we don't sweat. I do intend to follow this advice to the letter – but after only a few minutes, we're all sweating. We rummage in the badly lit and awkward corners of this tight space, trying to find our winter survival clothes. Our names are written on each item: a light suit and jacket, a jumper, a heavy suit and jacket. Then there are gloves, shoes and a hat. We take turns helping each other to doff the Sokol suits we donned this morning after the medical check-up, and we put on the first layer of thermal clothing. Then we finish getting dressed outside, where there's more room. After all, it's not particularly cold today.

To be honest, we're very lucky. The temperature is -10° C, and it may fall to -15° C overnight, with very little wind. In these conditions the snow, which comes up to our knees, will stay very dry even in the sun. Over the past few weeks I've followed the changing weather forecast closely, taking comfort in realizing that I wouldn't join the ranks of those astronauts who can boast of having survived at -30° C, with snow up to their chests. Those are great stories to tell after the event, but I'm sensitive to the cold and I find the Russian winter is quite intimidating.

Even in relatively mild temperatures, the next two days are going to be somewhat challenging. The scenario is actually rather improbable, especially since the Soyuz is now equipped with a satellite phone and a GPS receiver. We are simulating an emergency return to Earth, after which the search and rescue team won't immediately find us, and we'll have to manage on our own, using our winter clothing, the survival kit, the parachute and its valuable cords, and whatever else we can cannibalize from the descent module. The seat liners, for example, can be used like sleds for dragging things over the snow; or the Forel dry-suits from the water survival gear: you can cut off their legs below the knee to make high, waterproof boots.

By the time we've collected the survival gear and the parachute and we're ready to set out through the forest, there are about four hours of light left. We don't need to source food, since we have a few supplies with us, nor do we have to worry about dangerous animals, because the forest we're in is near Star City, and actually encircled by a wall. We do, however, have to work quickly to prepare a shelter and a signalling fire, as well as gather wood for a campfire before night falls. Sergei finds a good area: there are two trees about 2 metres apart, an ideal spot for building a quick shelter, and there's lots of space in front of them for putting up the more comfortable tepee tomorrow, on ground that will be warmed by tonight's campfire. We know very well what to do, since before Christmas we came to the forest with our instructors for a training session in preparing shelters and using survival gear.

I won't pretend I haven't often wondered how much we'd really be able to do if we had just returned from six months in space, when your muscles, heart and brain struggle to readapt to the burden of weight, and your vestibular system has to learn how to maintain balance. Of course, we all know that human beings make a virtue of necessity and can draw on surprising resources in an emergency. And in any case, whatever the physical limitations, it's always better if you know what to do rather than having to improvise. As it happens, as well as teaching practical skills, the survival course is considered an element of BHP (Behavioural Health and Performance) training, a chance to confront your reactions in uncomfortable conditions such as fatigue and cold, and find your own strategies so you can continue making positive contributions to the team. With a real Soyuz crew, it's also a great opportunity to get to know each other's strengths and weaknesses while cementing your bond.

We haven't been assigned to a space mission; we're a crew only for these three days. Thomas and I are here because ESA is finding us regular training opportunities, in my case as a reserve astronaut. Sergei is doing the course as a part of his basic training. Since he's a cosmonaut, he acts as commander for our simulated crew, and once he's found the campsite, he leads the activities and assigns duties. I immediately appreciate his leadership for being clear and firm, yet exercised with disarming grace. A former pilot of TU-160 'Blackjack' strategic supersonic bombers, Sergei has great outdoor skills, along with a natural instinct to care for others' needs. This all helps to establish a calm, easy atmosphere from the moment we begin to work together.

Sharing a knife and a machete from the survival kit, we cut down medium-sized branches and we use those and the parachute cords to build a frame for our shelter. Then we fix branches and leaves over the roof and spread them over the ground. Finally, we add two more layers on the inside: the parachute fabric and the reflective survival blanket. It's no palace, but we'll manage. When the sun sets, we'll be ready for the sixteen hours of darkness ahead. We've gathered enough wood for the night and we've built a big signalling fire. If we should get a radio call or hear a nearby rescue helicopter, we'd be able to start it quickly to indicate our position.

I'm really glad that work is finished for today, chiefly because I can now warm up around the campfire. Despite the fact that the temperature is relatively mild, I've suffered from the cold, mostly in my fingers, and the old-fashioned gloves from the Soyuz survival gear offer little protection. We spend the evening chatting, eating a little something and making sure the fire dries the wood for the night. After a few hours we go to sleep, taking it in turns to watch the fire and to make three mayday calls in the blind on the hour every hour, at precise two-minute intervals.

In the morning I wake up with stiff limbs from the cold and our Spartan bed of leaves and branches, and I feel the cold all the more since our meagre food supplies have left me hungry and I hardly slept. Still, I wasn't awake the whole night: I managed to sleep a little, in twenty- or thirty-minute snatches. Today we have eight solid hours of light, enough time to build a comfortable tepee. When we're finished, we light a fire inside, and the smoke escapes at the top, from the central opening, while fresh air is pulled in through the gap between the two strips of parachute fabric that we wrapped around the frame made of tall tree trunks.

That evening we receive a response to our mayday call. As expected, someone has found us. And as expected, our rescuers ask us to light a signal flare – there's one in the survival kit – and the big fire we've prepared. Finally – as expected – they tell us they've identified our position and will come back to collect us tomorrow, when there's sunlight. We'll have to be ready to leave our camp. They'll tell us which direction to go in to reach the rescue area, where the virtual helicopter will land.

The night is still young and the sun will rise only a little before ten in the morning. So here we are in our tepee, sitting out an exercise that is now mainly one in practising patience and putting up with the cold. To tell the truth, it's not that bad in the tepee, even if we should have made it a little smaller so it wouldn't lose so much heat. We sit around the fire in silence, enjoying the peace of the evening, and when it is time to turn in, I offer to take the first shift. Right now I don't feel tired. I'm enjoying the peace and the mesmerizing dance of the flames. After an hour, I wake Thomas and hand him the radio. Before slipping into a light sleep, I think about the steamy sauna we'll be able to take tomorrow – after we've survived the Russian winter.

—

When I got back to Star City in mid-November after that first intense month in Houston, I picked up my training where I'd left off in the summer, continuing the course on the Soyuz life support system. It's a combination of interdependent elements and functions, which include regulating the internal cabin pressure and the oxygen partial pressure, the removal of carbon dioxide, the drinking water dispenser and rudimentary toilet, and the Sokol suit. When it comes to preserving human life on board, there are many different issues. What happens, for example, if the partial pressure of oxygen drops too low, simply because the crew uses it up by breathing? Nothing too serious: a valve opens automatically, and extra oxygen from the tanks is released into the descent module. And what if the automatic system breaks down? If the oxygen partial pressure drops below 120 mm of mercury, an alarm and a warning light will alert the crew, who will open the valve manually. But it's important to close it at the right moment! Too much oxygen can cause a fire.

I had already learned this and a lot more the previous summer, in a number of theoretical lessons with Viktor, a middle-aged instructor with irresistible enthusiasm. He taught me, day after day, how to find my bearings in the schematic diagram on the wall. Baton in hand, he would point to pipes, valves, sensors, warning lights and fans as well as the underlying logic between them. The theoretical lessons were held in a room adjacent to the simulator hall that also opened onto the equipment room where many Sokol suits from past missions are kept; they are now used for training. Another Viktor presided over the equipment room. He was older but no less industrious, and he often helped the cosmonauts as they got dressed in their Sokols before a simulator session. 'Viktor and Viktor' formed one of the most likeable instructor teams in the whole of the ISS community.

I'll never forget the younger Viktor's zeal and total lack of embarrassment when he showed me how the toilet works on the very first day after my return that autumn. We climbed up the steep spiral stairway that winds around the simulator and leads to a small round hatch in the side of the orbital module. It really exists, that hatch: that's how cosmonauts would go out on extravehicular activities in the past. Today it's used to enter the Soyuz on the launchpad, but once the crew have taken up their seats, it's closed by ground personnel, never to be opened again. In the simulator, however, it was the access for our regular classes in the orbital module.

As always, Viktor and I left our shoes outside and went in on all fours, straightening up once we got into the small, spherical area, barely 2 metres in diameter. We shared a narrow walkway, just a strip, really, with a protective grille over the circular hatch which opens onto the descent module below. Years later, I was surprised to discover how much cargo will fit in that cramped area, and how little empty space there is during the trip to the ISS.

At that time, the journey lasted two days rather than the six hours it took me three years later. At that very moment, as Viktor was pointing out the life support features to be found in the orbital module, a crew up there were probably negotiating use of the small space with massive cargo bags. The previous night, in fact, Soyuz TMA-22 had been launched, swallowed by a blizzard only a few seconds after lifting off from a launchpad lashed by a violent snowstorm. On board were three genial colleagues, all of them on their first Soyuz flight: cosmonauts Anton Shkaplerov and Anatoli Ivanishin, former MiG-29 pilots, and NASA astronaut Dan Burbank, a US Coast Guard officer. I'd made friends with Anton over the summer, encouraged by his cheerful, sociable nature. I'd invited him to supper once when he was in Cologne for his last training session at EAC, and with a plate of lasagne in front of him – cooked by an eternal novice! – he proposed a toast as a courtesy to the host: we toasted in hopes of my first flight and his second flight coming soon, and to an assignment on the same crew. I might have saved that bottle of red if I'd known how prophetic our toast was to be.

The internal architecture of the Russian vehicles has changed very little over the course of decades and compared to the more sterile look of the western vehicles, it has a character I'd almost call homely. On one side of the orbital module there's a sort of bench covered in bottle-green carpet where you can attach Velcro. With Earth's gravity, and considering the small space you can actually stand up in, you naturally want to sit down on it, so much so that the Russians call the structure 'the sofa'. I perched on it, leaving space for Viktor to show me the features kept on the opposite side, behind removable yellow, ivory and pale-green panels. Hidden in there, for example, are cartridges of lithium hydroxide, a substance that reacts with the carbon dioxide produced by the crew's breath and traps it in a white salt. Behind another panel there's a container that collects the condensation in the cabin so that it won't cause excess humidity, which besides being uncomfortable, can steam up one's helmet.

The 'luxurious' toilet on the Soyuz is located in a tiny niche behind a curtain. I watched Viktor take a yellow plastic cone-shaped funnel from the wall. It was attached to a flexible hose and a little metal cup that looked like a watering can, and that in turn was hinged to the yellow funnel so that, when not in use, it wouldn't get in the way. I must admit: I struggled not to laugh while Viktor illustrated how to go with his usual enthusiasm and in great detail, without giving any sign that he realized the comic aspect of the situation – not even when he demonstrated how the cup swivels – if necessary – to fit against your rear end, while keeping the yellow funnel in the right position to capture the flow of urine. All of this while completely clothed, you understand.

As with all toilets in space, even the simple one on the Soyuz must satisfy two fundamental requirements: getting things to go in the right direction in an environment where everything floats; and what to do with the unpleasant by-products. The first issue is resolved by means of a fan that generates a flow of air and conveys liquid and solid waste towards their destination. Urine travels through the flexible hose to a container filled with absorbent material and faeces go into the cup, lined with a single-use insert that is eventually tied and put into two other pouches before being thrown in a large waste bag. Viktor spared no detail, recommending for example that we squeeze the air out of the insert before sealing it, and taking particular care to direct the smelly air flow towards the urine funnel so that it is sucked down the hose and through the filter, which helps to deal with the odour. Let's be clear: there's very little privacy on the Soyuz. If someone needs to use the corner toilet, the other two crew members go into the descent module and pull the hatch behind them, leaving it only slightly ajar.

Autumn was uncharacteristically mild, with hardly any snow, and it went by quickly, punctuated by exams and cheerful social activities in Star City. I couldn't always get together with the others for evenings in the *banya* or weekend outings to Moscow, since my difficult theoretical exams were coming thick and fast. I was happy enough if I could have a chat in Shep's Bar after supper. On Thanksgiving Day, the level of suppers in Cottage 3 went up a notch. We celebrated in grand style, with the traditional turkeys prepared by the Americans in Star City, who invited the Russian staff working for NASA to join the feast with their families.

The days were growing shorter, and I soon found myself setting off for classes in the dark, despite a start time of 9:00 a.m. The theoretical part of my flight engineer training was fast approaching its end. Complex courses, such as the one on the Soyuz re-entry system, alternated with easier short classes, illustrating, for example, the launch escape system, or how parachutes and retrorockets work during landing. It was almost Christmas when I took the last of the four formidable tests on the guidance, navigation and control system, this one focused on the rendezvous with the ISS, with all the thousand possible failures and back-up modes.

I was relieved to do well in the exam. I went back to my room in the Profilaktorium and took down the sheets of paper I'd stuck on the wardrobe a few days earlier – with hand-drawn graphics I had used to try to visualize the logic of a system that at first seemed to have none. I quickly packed my bags while Nikolai nervously waited for me, impatient to leave. Winter had finally come to Moscow, and the first real snow of this strange December had arrived with terrible timing today, when I had to tackle the unpredictable trip to the airport. In order to drive the 80 kilometres to get there, we were ready to leave five hours before the plane took off, and yet Nikolai felt that we were already behind. Traffic on the Moscow ring road is terrible even on normal days, so imagine what it's like when the radio announces that, on a scale from one to ten, the traffic level is nine. I hugged Yuri and Anna and wished them a Happy New Year and a Merry Christmas in that order, following the Orthodox calendar. Then my thoughts flew to Kazakhstan, where, in a few hours, my ESA colleague André, NASA astronaut Don Pettit and Russian commander Oleg Kononenko would be leaving on their Soyuz. At that moment they were probably putting on their Sokol suits for a final leak check before heading for the launchpad. In the frenzy of exams and preparations for departure, I'd forgotten to charge my phone. Although it was unlikely, I still hoped to arrive at the airport in time to watch the launch live.

After a couple of painfully slow hours on the road, the radio announced that the traffic level in Moscow was now measuring six out of ten points. I looked at the road and figured that all six of those points had to be due to traffic jams occurring on our route to the airport. The crew in Baikonur would already have taken their places in the descent module 50 metres up the rocket. Who could say what those three veterans were thinking when the hatch closed over them? I imagined that, more than anything, it must have been a great relief: they could finally focus on just one thing – the launch – and were no longer available for the thousands of requests, all of them well-meaning but as a whole tiring, that bombard any crew in its last few months on Earth. It had to be satisfying to be able to concentrate on the present. A bit of apprehension was inevitable, I thought, but the calm routine of pre-launch procedures, repeated so often in the simulator, would certainly be a solid anchor for the mind to settle on in that limbo between the comfort of the familiar and the uncertainties of the imminent voyage. Or so I thought. I didn't make it in time to see the launch, but I did see the final moments of the ascent, and I rejoiced along with my friends when I saw them join hands in celebration, acknowledging the shutdown of the third stage and the beginning of six months of weightlessness.

In January, after spending New Year's Eve in the Martian landscape of Wadi Rum, Jordan, I went back to Star City to find it well and truly wintry. I continued to use the bike; Yuri Petrovich had studded its tyres with nails. It was actually the safest way to get around, definitely less risky than walking on the perpetually icy roads. I tempted fate one day by taking a pot of ratatouille from the Profilaktorium to Cottage 3 and miraculously managed to cross the large court without falling.

When I left again at the beginning of February, I'd finished the complete Soyuz flight engineer course. If only I could one day be assigned to the left seat – I'd wanted it so badly for so long, and all the more so after all that studying – I'd be in an enviable position. With my theoretical course under my belt, I was ready to begin training in the simulator with the commander. Hopefully not too many years would go by before then, since the qualifications have an expiry date. To put it another way: if only I could fly on the 2015 ASI mission! By then, it could only be a few more months until the assignment was made.

10

'The Answer to the Great Question . . .'
'Yes . . . !'
'Of Life, the Universe and Everything . . .'
    said Deep Thought.
'Yes . . . !'
'Is . . .' said Deep Thought, and paused.
'Yes . . . !'
'Is . . .'
'Yes . . . !!! . . . ?'
'Forty-two,' said Deep Thought, with infinite
    majesty and calm.
        Douglas Adams, *The Hitchhiker's Guide to the Galaxy*

Houston, 5 March 2012

'EV1,' the test director, TD, calls out in a dry tone. A heavily built man in his forties, he's head of operations today. 'Go!' Tracy replies straightaway, just like all the others in charge of various aspects of today's training, when asked whether their respective areas of expertise are ready. They've all given the OK, from the doctor to the chief tool technician, from the lead diver to James, our instructor. In NBL jargon, James is called TC, or test conductor, while Tracy and I are the test subjects, or EV1 and EV2. When TD calls out 'EV2', I too respond with a decisive 'Go!', which shows my determination yet carefully conceals my apprehension – or so I hope. After long fit checks and countless classes, the long-awaited day has finally arrived. I am about to do my first suited run in the NBL pool.

The control room quickly empties after the briefing. There are still ten minutes to go before 8:30, when Tracy and I are due to be suited up. From the large window we can see the pool bustling with activity. The divers are preparing their equipment, and the technicians are almost finished getting the tools and the suits ready. Our suit torsos are already attached by their backs to the donning stand, on a platform that will be raised and then lowered into the water by a crane when Tracy and I are ready. Mini-workstations sit on a nearby table and will be installed on our suits once the suit-up and pressurization are complete.

We configured our mini-workstations this morning as soon as we arrived. I got here at 6:30 a.m. and Tracy arrived at the more sensible hour of 7:00 a.m. At 7:30 we were directed to the dressing room for a quick medical check-up before we put on the under-wear for spacewalks. First of all, there's the MAG, or maximum absorbency garment, nothing more than an adult nappy. It's rare to have to use it in the pool, I was told, but it is reassuring to wear it, all the more so in space, where the astronauts spend up to twelve hours in their suits. I put on simple, white cotton leg-gings over the MAG, socks and a long-sleeved undergarment, all of which protect the body from abrasion. Lastly, I put on the LCVG, or Liquid Cooling and Ventilation Garment, taking care not to damage it. This is a white unitard that adheres tightly to the body, leaving only hands and head bare. It incorporates 80 metres of flexible transparent tubing through which cooling water flows to remove excessive heat from the body. Here in the pool, where we don't have the real autonomous suit life support system, the water comes from the NBL plumbing through the umbilical, a bundle of hoses and cables that also provide the Nitrox feed for breathing and pressurization and the audio connection.

As I come out of the control room with Tracy and head for the suit-up, I nibble a high-calorie peanut butter bar, somewhat against my will, since I'm still full from the eggs and porridge I had at breakfast. All the same, I force myself to eat this snack because there's no food in the suit, and I'm a little obsessed by the idea that I might not have enough energy. I've lost count of how many people have warned me that working in the suit is exhausting.

Actually, hunger shouldn't be a problem today. This first

session will last for only three hours, and then the hours will be progressively increased to the full six by the time we get to the fourth run. We can take our time suiting up, and Tracy takes the opportunity to show me the tricks of the trade. She speaks as an EVA veteran, having made three contingency spacewalks in orbit to replace one of the large cooling pumps of the Space Station, which had broken down.

I find that a few of my friends have gathered around the pool to share this special day with me. There's David, the Canadian doctor and astrophysicist from our twin class, the Chumps. The undisputed world champion of anything to do with hospitality and care, he immediately offers to document the event with my camera. I remember that I need to give back his ski gloves, which I asked to borrow for my visit to the tools lab so that I could simulate, at least in part, the difficulties faced in wearing the real ones with the suit while I practised and gained confidence with the tools. Luca is also here. He's been training in the NBL for more than a year with excellent results and has generously shared the techniques he's learned, even though we both know that his size is ideal for the suit, and allows him to do things that will prove impossible for me. And naturally there's Cady, and she's brought her eleven-year-old son, a bright and sensible boy who chants a light-hearted 'No pressure' to encourage me to enjoy the day.

Tracy, too, has repeatedly made it clear: today is for me, and belongs to me alone. No one is expecting anything of me. I just have to become familiar with the suit underwater, maybe even have fun with it. I like the idea, but I know very well that it's not entirely true. Even though there won't be any formal evaluation, it's surely desirable to leave a good first impression. Each and every person at the NBL will form an opinion of me today.

I sit on the ground on a mat and put my legs into the legs of the suit, pointing my toes to burrow my way little by little through the folds of the inner yellow membrane. Grabbing one hand each from two technicians, I stand up and then crouch under the torso, which is already hanging at the donning stand at just the right height so that once I'm in it, I can remain standing

88

on the platform. With a sort of squat I push myself up into it: first one arm, then another, until my head and hands pop out of it. Something doesn't feel quite right on my back, around my shoulder blades. Maybe it's just normal discomfort; after all, the EMU isn't really meant to be comfortable. I have so little experience that I almost don't dare mention the problem. Luckily, however, Steven is here to supervise the suit-up process and right away he notices what's wrong. He puts a hand into the opening at the neck and finds the large stiff ventilation tubes, which must be overlapping. An integral part of the LCVG unitard, they run along the legs and the arms, joining up on the back. In space, they retrieve the used oxygen from the areas around your ankles and elbows – which is now mixed with carbon dioxide and water vapour – and take it back to the revitalization system. Here in the NBL, they retrieve Nitrox in the same way, and it is simply sent back to the surface through the umbilical and replaced by new gas, which is fed into the helmet.

With the connection closed at hip level, Steven and the technician each take one of my hands and stick small moleskin patches on them, prepared according to a detailed checklist devised during the glove fit check. These will protect my hands from potential abrasion. I put on my gloves, and then it's time for the helmet, which clicks shut over the metal neck ring – and the hermetic sealing of the suit is complete. Now we can pressurize. I'm wearing the Snoopy Cap and through the headphones I hear TD's voice reminding me that I can ask them to stop increasing the pressure at any time if I have problems popping my ears to equalize. I don't expect any difficulties. I don't have any cold symptoms, and during fit checks and the Prep and Post it was easy for me to perform the Valsalva manoeuvre using a simple device – a piece of soft rubber – glued at the base of the helmet: you press the base of your nostrils against it and you seal them. This way you create overpressure in your airways to balance the increased external pressure. You do the same thing when you go diving or your plane is descending: you squeeze your nostrils together and blow.

A little while ago, before my suit was pressurized, it felt like any item of clothing, if a bit odd. Now that it's stretching out and stiffening, cracking here and there, to me it looks more like what it actually is: a small spaceship, capable of keeping a human being alive in space. I have to confess, it also makes me think of those giant robots with human pilots on board in the Japanese anime series of my childhood. I've always been fascinated by the synergy between humans and machines, and the possibility of using technology to exceed the limits of the human body. Just as when I was training to be a military pilot, I'm attracted today by the challenge of learning to merge with a complex machine, deeply understanding it, knowing how to exploit its potential and work with its limitations. I like knowing that it isn't easy, that it's a long journey, that you gain the necessary skills only with a combination of sustained intellectual, physical and psychological effort. On the other hand, I am also thrilled to be on the verge of experiencing something unusual and exciting and – yes – not available to everyone. And just like the first time I flew a jet, part of my being is distilled into the pure and simple yearning for experience – as if, as a child, I'd been able to take Tetsuya's place in the head of the Great Mazinger.

After a few photos with my friends, the mini-workstation is installed, and at that point everyone is asked to move behind the yellow line so they won't get in the way of the crane. The platform is raised and moved over the water before being slowly lowered. I wave to my friends on the pool deck and just before my helmet is submerged, I close my eyes for a second, a trick I'm told will prevent my being disorientated by transitioning into the water. When I open them again, I'm stationary just below the surface, and the divers – happily, I see Sarah among them – are already busy around me. They detach me from the donning stand and shake me for a few seconds to free the air pockets trapped in the folds of the suit before checking to make sure there aren't any other bubbles, which would indicate a leak. When we get the green light from Steven, who's watching from the poolside, they drag me across the surface to the descent line on the other side of the pool. It's an interesting effect, this bird's-eye view of that intricate, submerged city. The mock-up of the Space Station is lying down there. I've explored it so many times by now during dives, but seeing it from inside the suit – when I'm totally relaxed and powerless in the hands of the divers – is completely different, almost surreal. If it weren't for my suit, I'd have to pinch myself.

Once we get to the bottom of the 13 metre-deep pool, the divers start working on the weigh-out. They work carefully, turning me round and round in every direction to observe remaining tendencies to rotate. Meanwhile, TD initiates a comm check. He instructs Tracy, and then me, to count slowly from one to ten, to confirm that we can hear each other and that TC can also hear us from the control room.

'TC, from TD, the test is yours,' comes through the headphones as soon as the weigh-out is completed. James, our instructor, is now in charge. My task for today starts now: I'll explore my work envelope, learn the limits of the EMU, get used to its size and my restricted field of view, practise moving and changing direction, identify any need to adjust the suit fit. It's not possible to bend or turn your arms beyond the limited range of motion allowed by the joints of the suit; if you force things, you risk damaging your own joints. There's no point trying to turn your head to look up or sideways: you have to turn your entire body. There's no way to move quickly; turning or moving from one place to another requires focused and conscious effort, to say nothing of patience.

A sequence of small exercises is planned for this first run. We start by practising translating at the easiest location possible, that is, along the vertical side of the truss, where there's continuous railing. From there we'll move down towards the Lab and Node 2, a translation that's more difficult since I have to overcome obstacles and change orientation more than once. Then I'll try using the PGT before the divers take me to an APFR, or Articulating Portable Foot Restraint, a device that allows you to lock the boots of your suit in order to work freely with your hands, yet with a solid anchor point. Getting your boots into the APFR, however, is no easy feat. Or to clarify, the movement itself isn't difficult – a deliberate rotation of your heels – but it's a huge challenge to orientate yourself correctly and to find the right handholds to press your feet onto the APFR plate. It would, of course, be much easier if I could see my feet, but that's a privilege reserved for tall people with long necks, whose heads fit higher inside the helmet.

Fortunately, I had no illusions. Every little thing I do in the suit feels extremely difficult and tiring, but I'm not frustrated. On the contrary. Maybe I expected worse. Apart from anything else, a few small worries I had have proven to be unfounded. For example, the drinking straw is well positioned, and I can take sips easily. It's like the straws used by cyclists: the bite valve opens up when you squeeze the straw between your lips, and then closes when you're done. The drink bag, which I attached inside the suit under Tracy's guidance, holds roughly a litre of water.

Towards the end of my three hours, they've reserved time for a couple of demonstrations. One consists of a ten-minute period with my head down, which is fairly unpleasant because of the pressure on my shoulders. Finally, when I'm back at the donning stand, and still just below the surface at the side of the pool, I'll have a chance to experience rapid depressurization. First, for safety reasons, the pressure of my suit is reduced, then a diver removes one of my gloves, and my suit immediately deflates. It's a situation I'll probably never face again, but I'm glad to know now that there's nothing dramatic about it.

I'm radiant when I come back up to the surface, beaming at everyone around me. You don't have days like this very often, where you experience something truly exceptional. Moreover, despite all the difficulties posed by my small size, I wasn't uncomfortable. I'm actually confident that with time and a lot of practice, the suit and I will find a way to get along.

Today, a bit more than usual, I feel like an astronaut.

—

It was one of the most thrilling experiences of my life, the kind you relish little by little because you got there slowly, one step at a time . . . so much so that when they actually do happen, they seem almost inevitable, and it's only later that you realize how exceptional they really were. For some time, I continued to relive those moments of my first day in the suit, as if hoping to keep the memory from fading away. Yet only a short time later, something happened that would trump everything else, emotionally speaking.

I don't really understand the chain of events that led up to it – the time difference between Europe and America no doubt had something to do with it. But the news came from Silke, secretary to my boss in Cologne, who forwarded a message sent a few hours earlier. The subject line read:

Fwd: Fw: 2 letters from ASI regarding the assignment of
Samantha Cristoforetti to Expedition 44/45 in May 2015.

The waiting had finally come to an end! No more worrying or uncertainty, no more fear of unforeseen events. Attached to the email were the official, signed letters from the president of ASI, recommending my assignment to NASA and ESA respectively. It was a watershed. For those who knew nothing about it, I was the same person I'd been the day before. But for me, everything had changed. The countdown had begun: in less than three years' time, I'd be going into space.

The news of my assignment ushered in a slew of questions, and I was particularly keen to know what seat I'd be assigned on the Soyuz. Would I be flying, as I hoped, in the left seat in the role of flight engineer? To find out, I would have to wait for the MCOP's formal appointment. The MCOP was a committee formed of all the international ISS partners, and together they would decide on the crew. Perhaps precisely because the seat assignment was still pending, the same day, Dmitri asked me to give him my availability for a retake of the Russian exam, since my original result, though perfectly good, was now two years old. After all the months I'd spent in Russia the previous year, and thanks to the excellent teacher who gave me occasional lessons at JSC, over the next few weeks I attained my highest level of Russian ever and got the best possible marks, surely a point in favour of the assignment I longed for. And it had to count for something that, as a reserve astronaut,

I'd already completed the theoretical part of the flight engineer course.

It was actually the MCOP that confirmed after its meeting on 4 April a persistent rumour that had been circulating for a couple of weeks, to the effect that the ASI flight was being brought forward by six months. I would be the flight engineer – hurrah! – for Soyuz TMA-15M, with an expected launch date of late November 2014. As if that weren't exciting enough, there was also a geeky aspect: my crew's arrival at the ISS would complete Expedition 42. As every reader of Douglas Adams and his *The Hitchhiker's Guide to the Galaxy* knows well, forty-two is the answer to the ultimate question of Life, the Universe and Everything. The exact question is never revealed in the book, but that's the beauty of it. As far as I was concerned, 'Which ISS expedition will you be on?' qualified without a doubt as the ultimate question.

With my launch date and my role on the Soyuz confirmed, my greatest curiosity concerned my travel companions. Who would they be? Veterans, or rookies like me? Men or women? Military or civilians? Pilots or scientists? A doctor? Would they be roughly my age, or older? From the MCOP, again, I had the answers to all my questions. Besides me, they'd assigned cosmonaut Sergey Zaletin and NASA astronaut Terry Virts, the same Terry I'd seen fly into space a couple of years earlier on the Space Shuttle. Our assignment also meant that we would serve as back-up crew for the Soyuz leaving six months ahead of ours. So Terry and I were now back-ups for my Shenanigan colleague Alex, a PhD in geophysics with a passion for active volcanos and wingsuit flying, and for Reid Wiseman, a NASA astronaut and former US Navy test pilot.

Our mission wouldn't be publicly announced for some months, and we were not allowed to release any information earlier, but the machine that would propel us towards the launchpad immediately swung into action. Things moved slowly initially, but they would accelerate, relentlessly up to the launch day. One of the little signs that things had changed was that I was assigned to Luca A., an increment training integrator, whose job was to schedule my training.

For the next two and a half years, my life would belong to him.

II.

> . . . but if something is revealed to me this evening, it
> will be because I have laboured to carry my stones to the
> invisible place of construction. I am making ready for a
> celebration. I shall have no right to speak of any sudden
> apparition in me of any other but me, for I am building
> that other.
>
> Antoine de Saint-Exupéry, *Flight to Arras*

Montreal, 13 May 2012

I've been in Montreal for a week, renting an apartment in the
charming Plateau Mont-Royal quarter. I haven't gone anywhere
besides work, with the exception of brief strolls in the neigh-
bourhood. I've really loved exploring the quaint tree-lined alley-
ways and terraced houses, each with an external staircase leading
to a separate entrance on the first floor. Straight or curved, plain
or fancy, all these metal staircases with open steps evoke a liveli-
ness that corresponds perfectly with the colourful energy of this
quarter.

After an early morning walk, I make my way each day to the
Canadian Space Agency's facilities for training on the SSRMS,
the huge robotic arm on the ISS. It's a two-week course, with
thorough theoretical lessons on engineering aspects alternating
with practical sessions at the simulator. The simulator is fitted
with a robotic workstation identical to those on board: there
are two hand controllers, which I already know well thanks to
the course I took in Cologne during my basic training; a computer and a control panel for configuring the system and sending commands; and three monitors showing images from the

video cameras. There's no real robotic arm, but the simulation software shows the exact images we'd see if we were at the controls of the SSRMS in orbit. On a table beside the workstation is a large and very detailed 3D-printed model of the ISS. I found it an object of great beauty, and any model-builder would go crazy for it, but above all it's a visual aid to prepare you for the mental gymnastics to follow: predicting how the movement of the arm will look from different viewpoints, identifying the best combination of video cameras, mentally reversing their images, deciding how to deflect the hand controllers in order to achieve a particular movement based on the selected coordinate system, visualizing how the configuration of the robotic arm changes during this movement . . . These are some of the tasks that keep your brain well occupied when you're in command of the SSRMS.

The robotic arm has no hand. In its place, or rather at both extremities, are two identical end effectors, both of them shaped like squat cylinders. With one of them the SSRMS is attached to the Space Station structure; with the other, it can grapple any object that is equipped with a grapple fixture. The object could be a large spare unit for the external machinery of the Space Station that needed to be relocated, or even an entire module, such as the Columbus, which was removed from the Space Shuttle cargo bay at the time of installation, and joined up with the rest of the ISS. Or it could be a cargo vehicle that had come to resupply the Space Station and needed to be 'captured' – so to speak – while it was holding position at a distance of 10 metres: the so-called Track and Capture.

For a few years now the SSRMS has been used to capture the Japanese HTV vehicle, and it will shortly start capturing the American Dragon and Cygnus vehicles too. As of next week I'll begin training for Track and Capture. It won't all be new to me, since the visual cues for aligning the end effector with the grapple fixture are exactly the same as those I've been using since my basic training. The difference is that with Track and Capture, the target is moving.

———

It was really lucky that my introductory course on the SSRMS, the only two weeks of training that take place in Canada, were scheduled for May, when the weather is mild and conditions are fairly pleasant overall. David, who was doing part of the course with me, took me to his family chalet set on a beautiful lake. I hadn't seen Andrey, a Russian friend who'd emigrated to Canada, since university in Moscow, and now I could go for a walk with him and his young family in the botanical gardens. And the instructors at the Canadian Space Agency invited me to a small gathering at one of their houses after work. I can still remember all the happy kids – at least twenty of them – in the garden, jumping on a huge trampoline. All this would have been impossible during Montreal's long, snowy winter. The course had been interesting, and I was leaving with a high score in the final exams, which boded well for the next six weeks of robotic training in Houston.

In terms of EVA, there'd been an unexpected development. Since I'd been assigned to a mission, I was now at the top of the priority list for custom-made gloves, and they'd immediately scheduled a morning for measuring and taking a mould of my hands. It would take a year to get the first prototype and overall a couple of years to have the final pair to use underwater, but I could already taste the joy of putting them on.

In the meantime, I'd finished the introductory cycle of four sessions in the pool, including the first one lasting six full hours. Well aware that my size put me at a disadvantage, I prepared for every run with maniacal attention to detail. By this time, Cady was used to seeing me at her kitchen table at odd hours, watching videos of my underwater work so I could find ways to improve. And before every pool run, I'd scuba-dive in the NBL to inspect the translation paths and the worksites, trying to memorize the position of obstacles and no-touch zones. It was easy to see them through a diver's mask, but just as easy to miss them in the suit, with the helmet's restricted field of view.

When I'd taken the '1G', the lesson that takes place before every run in the pool and in which the instructors point out the features and oddities of the components with which you'll be working, I spent my evenings and weekends writing and rewriting the procedures, trying to streamline the sequence of operations, the configuration of the tools or the 'tether plan' to avoid crossing or tangling the safety tethers. When I'd finalized the procedures, I performed meticulous 'chair flying', a technique borrowed from the piloting world, and involving a vivid and detailed visualization of a complex activity in its entirety or at its most critical phases. Of course, I wasn't preparing for flying sorties, as I once was, but this was the same concept: mentally training yourself to complete a series of operations planned in advance.

Since I wasn't part of a class of astronauts taking the course together, I usually found myself paired with veteran astronauts. Two of them actually offered to partner me and acted as informal instructors: Peggy Whitson, then chief of the Astronaut Office, and Sunita – called Suni – Williams, who was preparing for an imminent launch. I was amazed that such busy spacewalk veterans would take the time to train with me in the pool.

The introductory cycle was finished, and now the EVA Skills Course began: four more runs in the suit, to be followed by a qualifying exam. My days at the NBL began early. At around six, I'd pick up a couple of boxes of fresh *kolaches*, pastries of Czech origin which are very popular in Texas, and a couple of litres of coffee from the local Starbucks. It was tradition to leave it all in the NBL control room as a courtesy for the divers and the rest of the staff. I was usually already on the pool deck by 6:30, configuring the modular mini-workstation and the toolbags. After six hours underwater and a quick shower, I'd return to the control room for a debrief, going over things with the instructors, analysing, commenting and paying particular attention to areas for improvement.

After a day in the pool I usually went to bed early, exhausted, and the next day I'd feel an ache in muscles I never knew I had. My chin was often swollen, and it stuck out, probably because it rubbed against the neck ring of the suit. It didn't hurt, but for several days I had the characteristic profile of a fairy-tale witch. Even today, if I look carefully, it seems there's a certain asymmetry to my chin, as if the swelling on one side never entirely went down. Yet there was a great sense of satisfaction. Every training session at the NBL was a small peak to scale, and reaching the end of an expedition was always fulfilling. It felt like all my hard work was paying off, since I was getting moderately positive feedback. Of course, no one thought I'd become a star with EVAs, just as I'd never become a basketball or volleyball champion. But I didn't expect to encounter any problems with completing the course or obtaining the qualification. Alas, future events were to prove me wrong.

Alongside EVA and robotics, there were still dozens of courses to take on the ISS systems – less difficult, it's true, but they nevertheless filled my days and the frequent exams required some preparation. So I had very little time for social life or recreation, to the point where David probably lost track of how many times I turned down invitations to go sailing with him.

I left for Europe at the end of June, relieved to escape Houston's hellishly hot summer, as well as the freezing cold air-conditioned interiors, which forced me to wear fleeces and jumpers even in high summer. At the ASI headquarters in Rome, however, where the public announcement of my assignment took place in early July, I reached the conclusion that too much air-conditioning is still preferable to none. I took a few days' holiday in the Alps and made a brief visit to Col-CC to meet the flight control team assigned to my ISS expedition. Soon afterwards, I began training again, moving from Cologne to Japan, then heading east to the US once more, and finally returning to Europe several weeks later.

'Slow orbit': that's what astronauts call this round-the-world tour. I would make it more than once during my years as an apprentice astronaut.

12.

An expert is a person who has made all the mistakes that can be made in a very narrow field.

    Attributed to Niels Bohr

Houston, 29 August 2012

Five minutes. I don't believe it! Five wretched minutes. I had two and a half hours to complete the tasks and return to the Airlock, and it took me precisely two hours and thirty-five minutes. I have failed the EVA Skills exam.

If I'm honest, I thought I'd go over time even more, and in that case, maybe I'd have been less upset with myself. During the final stages I was already convinced that there was no way to finish in time; if I hadn't given up too early, I might not have wasted those five minutes.

I don't understand how it could have taken me so long. After all, the tasks were not that difficult. There was very little in the Airlock, for example, apart from me and Serena in our bulky suits – just a small toolbag for each of us. In some of the past runs, the Airlock was packed with much larger bags, and more of them – and yet only today, for the first time, did I get a horrible tether snag. It took us quite some time to disentangle it, and things went from bad to worse from then on.

It's not over yet. As part of the exam I still need to perform a simulated incapacitated crew-member rescue. This scenario has never occurred in orbit, but you have to be prepared. You always go out in pairs for an EVA, so you can count on mutual support in case of emergency. At the NBL we simulate the worst-case scenario, involving a colleague who has lost consciousness and is therefore completely unable to assist in his or her rescue. The requirement is to get the incapacitated astronaut back inside the Airlock from any point on the ISS within thirty minutes.

Though the overall grading is already clear, the rescue goes ahead. The divers position Serena on the Columbus mock-up at one of the furthest points from the Airlock, in the ISS's usual direction of flight. Her job is to remain entirely passive, as if she's lost consciousness. When I get to her, she's floating in the water, attached to the structure with one of her local tethers, which are about 1 metre long. I release it in order to take Serena with me, but only after having secured her suit to mine. I also remove her safety tether reel: I have my own, and having two would only increase the risk of getting caught.

I start for the Airlock, now pulling Serena by the tether that joins us, now grabbing her mini-workstation to stabilize her and give her a push in the right direction. Everything is difficult in the EMU, but the rescue is on another level of difficulty altogether. We're not in the vacuum of space: the water offers resistance to the massive suit I'm dragging. I push and I pull, taking care not to run it into the structure, especially not with the fragile visor. It's also difficult to manoeuvre Serena inside the Airlock, and then to get inside when it's my turn and she's unable to help, at the very least by actively making space inside that confined compartment. The Airlock operations are always the most difficult and exhausting. Today, though, everything seems more taxing than usual. Maybe it's just my tiredness and disappointment, but there seems to be some anomaly, something in my way as I try to force Serena inside. Finally I do, but when it's my turn to enter, I find it unexpectedly difficult to get myself into the right position. When I'm in at last, reaching the button that releases the hatch is incredibly strenuous. By this point it feels like I've exhausted any remaining strength in my hands.

Ultimately, I manage to complete the exercise with a few minutes remaining but it's clear that this part, too, failed to go as it should. What happened exactly, I'll surely soon discover in the debrief.

—

I found out that while I was carrying Serena to the Airlock, the tether that joined us had wound between my legs and then hindered my later actions, limiting my ability to move. To the divers and TC, watching the live video, it was clear from the start. But I had no idea. I learned a lesson: if you sense there's something wrong, it's probably because something *is* wrong. It's pointless to struggle and proceed through sheer force of will. It's much more intelligent and efficient to stop, analyse the situation, resolve the problem and go on – even, and most especially, if you are in a hurry.

As far as the overall exam was concerned, I was terribly let down. I'd worked so hard and I'd hoped for quite another result. I wasn't sure what this hiccup would mean. I knew about others who'd failed the EVA Skills exam and had repeated the whole cycle of four runs before trying again. The difference was that they had done the course during their basic training. As an astronaut already assigned to a mission and bouncing from one continent to another, I found the idea daunting.

The hotshot in me would have been right back in the pool just to get it over and done with. Well, maybe not immediately – better to give it a few days so my hands and forearms could recuperate. The thing was, I'd never had serious problems and I'd missed the target by only five minutes; if I'd been able to repeat it, I'd surely have succeeded. My more sensible side, however, accepted that the exam had revealed gaps in my preparation. The tether snag in the Airlock, for example. In a sense, it was unfortunate that it hadn't happened before, perhaps because I had almost always trained with very experienced partners. Without even realizing it, simply by following their good habit patterns, developed over years of training, they had probably provided for good tether management. The best training allows you to make all the errors you possibly can, and I hadn't yet experienced that one. On the whole, it was clear to me that I had to improve my efficiency: I did things pretty well and safely, but I had to be much faster. In short, there were many things to improve and I didn't mind having a few more additional runs, though facing another exam was an unwelcome setback.

I really hated being a problem for the EVA community. In the coming days, there would no doubt be dozens of emails on the topic: informing, discussing and deciding on the plan of action, and then they'd let me know. I passionately wanted to be able to do an EVA in space, and this wasn't the sort of attention I wanted to generate.

If nothing else, I could console myself that I'd got an excellent result in robotics. After the two introductory weeks in Canada, I'd carried on with SSRMS training in Houston, taking the Specialist Skills course, a very intense programme of simulator sessions focusing on the two main uses of the robotic arm: in addition to Track and Capture, which relies on visuomotor skills, there's the more cerebral EVR.

EVR means robotic support for spacewalks. Sometimes it involves moving a large spare unit from one point to another on the ISS; for example, removing a broken-down pump after the astronauts outside have detached cables and hoses and unscrewed the bolts. Sometimes it's necessary to move an actual astronaut, who will have installed an APFR on the end of the SSRMS and secured their boots in it. The coordination is extremely tricky, since in some cases the person at the controls of the SSRMS is receiving instructions on how to move the arm from colleagues on the outside, based on their positioning needs. Responsibility for avoiding a collision with the Space Station, however, remains with the robotic operator, who has to know when to stop. Track and Capture, too, requires some coordination, because some monitoring tasks are carried out by the second operator, M2. Yet at the core of this activity is manual skill at the controls of the SSRMS: you need to compensate for the movements of the cargo vehicle being captured, maintaining the right speed of approach and aligning with the moving target, while avoiding any abrupt command input that might induce oscillations.

The exam took five hours and was quite formal, involving several evaluators and an experienced astronaut. I got a top score overall – M1 strong – which would allow me to carry out all types of robotic activity on board. What's more, the ISS was about to become a much-visited port: the Dragon cargo vehicle had just reached it for the first time, restoring the capability to bring useful cargo back to Earth that had almost disappeared with the retirement of the Space Shuttle – the only cargo return option being the very little space on the Soyuz descent module. And the following year, the Cygnus cargo vehicle would begin service.

After a couple of weeks in Europe, I set out for Alaska in early September, for a unique training programme: seven days in a kayak with Terry. This outdoor leadership course was suggested by NASA as an opportunity for us to get to know each other and to reflect and work on our expeditionary behaviour, those good practices equally suitable for a team adventure in nature or in space. There are other, more specific courses in so-called analogue environments, places and contexts that share some of the features of life on the Space Station. NEEMO, for example, takes place in Aquarius, an ocean-floor habitat off the coast of Florida, while ESA organizes CAVES in Sardinian caverns. At the time, however, the course in Alaska was better suited to my needs, since some of the other participants would be on my spaceflight crew. We were a group of eight. Besides Terry, there was Barry Wilmore – everyone called him Butch. A former US Navy test pilot with more than 6,000 flight hours under his belt and hundreds of carrier landings, he was part of Expedition 41/42 and would be leaving for the Space Station a few months before Terry and me. He would welcome us as commander when we joined Expedition 42, along with his Russian Soyuz crewmates Yelena Serova and Sasha Samokutyayev. When they left, Butch would hand over command to Terry, commander of Expedition 43, and we in our turn would welcome three new crew members a couple of weeks later.

The course in Alaska was designed to put us in a situation of fatigue and discomfort in which we would have to rely on each other to ensure the safety and well-being of the group. We would also have to take turns in the role of leader, with our guides and instructors on hand to offer suggestions and feedback on our performance leading the group. As far as I knew, the fatigue was meant to be mostly physical, and caused by the requirement of covering many kilometres per day in the kayak, putting up and taking down camp every time. But the bad weather in Alaska had something else planned for us entirely. The sea was so rough that we had to stay at the same camp, in the same place, almost every day. Our faithful companions were temperatures just above zero and the never-ending rain, and there was only one afternoon when the sky cleared enough to reveal the beauty of Prince William Strait. Each day we had to weigh the risk of continuing our expedition, because the radio was broadcasting warnings about approaching hurricane-force winds.

All told, we didn't get that tired, considering all the disruptions to our journey, but there was plenty of discomfort, and it gave us the opportunity to demonstrate our tolerance in the face of adversity, something that's always reassuring to see in yourself and your travel companions. It's equally important to look for ways to contribute to a group's well-being. Whether it was going off to fill water bottles, leaving the shelter and braving the pounding rain to move a tent that was on the verge of being flooded, or just doing the washing-up, someone was always offering to do even the most unwelcome jobs. All the inactivity in fact gave us a lot of time for talking in the communal tent, so much so that by the end we knew more about each other than we would have if the weather had behaved. I watched Butch and Terry with particular interest. They were going to be my two commanders, responsible for managing life on board, shaping the group dynamic and carrying out crucial communications with Houston. The two of them had very different leadership styles. Butch was infectiously cheerful, with natural charisma and a tendency to take quick decisions, or at least offer a solution he'd already worked out. Terry, on the other hand, was much more cautious, curious about everything and inclined to make joint decisions. No one form of leadership is better than the other in the abstract, and certainly both of them would adapt as necessary to the circumstances. Overall, I felt that our crew would be in good hands.

I returned to Europe completely restored by those ten days in the open air, and soon I left for Star City, where as a back-up for my colleague, Alex, I would have an unexpected opportunity. That autumn, I would train for the Russian spacewalks in the Orlan suit.

13.

Moscow, 19 October 2012

The procedures are written out on large sheets of cardboard, bound by a ring and tied with string to the frame supporting the pneumatic interface and the EVA control panel, which is peppered with command buttons and lights. A large blue knob in the centre of the pneumatic interface opens and closes oxygen lines to the suit and is used for choosing different pressure settings, depending on the various work phases. In orbit there are two identical interfaces, one in each of the small modules that serve as airlocks for Russian spacewalks.

It's Alex who reads out the procedures now, while I activate buttons and valves as necessary, and according to his instructions. First of all, we perform a leak check, opening the oxygen lines to pressurize the suit. We watch the pressure gauge carefully for around one minute. There is some slight movement, but the drop in pressure is within acceptable limits. I've performed this check in Star City several times over the course of the past few weeks, but today I pay particular attention. After all, on those earlier occasions, no one intended or had the means to remove the air around us. Today, though, here at the premises of the spacesuit manufacturer Zvezda, that is precisely the plan: Alex and I are in a big vacuum chamber.

In some ways, today's drill is not very different from previous simulations. We arrived in a van this morning from Star City, after the usual traffic ordeal, and were taken to a little changing room where we put on our bright-blue cooling undergarment. It's like NASA's LCVG, a unitard with the cooling water tubes woven in the fabric, but it comes with a hood and seems less fragile. In this get-up, which would seem bizarre in most places, but certainly isn't here in Zvezda, we were taken down long, gloomy corridors with brown laminate walls and printed linoleum flooring, until we finally emerged into a large hall with high ceilings, dominated by a sort of metal whale with its mouth open: the vacuum chamber. It's a large white cylinder about 4 metres in diameter and 7 or 8 metres long. When we arrived, it was closed at the far end and open at the near end, like a jar with the lid off. The latter was hanging from a crane a few metres away.

In the belly of the whale, which is covered in grey metal plating, our suits were waiting, hanging from a suspension system. Like the one in Star City, this system allows us to move our 150 kg of weight fairly easily. After a quick safety briefing and the reassurance that it's possible to repressurize the chamber rapidly in case of emergency, we turned on the power and, as our Russian colleagues love to say, we entered the suits. Indeed, you don't 'put on' the Orlan, you get inside it: it's a single piece, apart from the gloves. Just like on the EMU, the life support systems are in a sort of backpack, but on the Orlan the backpack is hinged and can open like a door. You put your legs in the suit while sitting at the door, and you connect the cooling water hoses and the cables for audio communication and medical telemetry. You can then turn on the fan and the water pump before you finally slide inside, letting your feet find the boots and stretching out your arms until your hands fit inside the gloves. These are quick and simple operations, especially when compared with the complex EMU-donning procedures.

When they were sure we'd got into our suits correctly, the specialists left, and I imagined the large lid behind us beginning to creep up, slowly, slowly, to seal the enormous jar. What we're doing today is a kind of final trial. We've passed a very detailed theoretical exam on the Orlan systems and now we have to demonstrate that we know how to carry out all the operations pre- and post-EVA. When this phase is complete – and only then – we can start training with the suit underwater – a very different approach to that used in Houston, where everything happens in parallel.

After the leak check, we re-equalize the pressure with the outside, letting the Orlan deflate. Later, when the pressure around us decreases to almost zero, the suit's regulator will intervene to maintain an internal pressure of 0.4 bars, or 40 per cent of the atmospheric pressure at sea level. If we were to breathe normal air in those conditions, the partial pressure of oxygen, which is only 21 per cent of the mixture, would be too low to keep us alive. So today it's mandatory to replace all the air with oxygen, just as we would before a genuine EVA in orbit. We open the oxygen lines in purge mode and wait for the air to be forced out of the suit.

Dropping down to 0.4 bars of pressure can bring on another problem, which scuba-divers know well: the risk of the bends, or decompression sickness. In our body's tissues there's always a certain amount of nitrogen, an inert gas that makes up 78 per cent of the air we breathe. If the surrounding pressure increases, as it does during diving, more nitrogen diffuses into your blood and tissues. This extra nitrogen is released both during and after the ascent, in much the same way that carbon dioxide bubbles up in a sparkling drink when it's opened. To ensure that the gas bubbles released are not too big, and therefore dangerous, divers make decompression stops during ascent: the longer and deeper the dive, the more stops they make. This allows the nitrogen to diffuse gradually from the tissues into the blood, where it is carried to the lungs safely. Going out in the Orlan suit at 0.4 bars is similar to an ascent, since the surrounding pressure decreases. The solution in this case is not a series of decompression stops, but a preventative desaturation: during what we call pre-breathe, we breathe pure oxygen for half an hour, and this induces a gradual diffusion of nitrogen out of our bodies. It's another step in the procedure which is usually only simulated, but today we actually carry it out. After thirty minutes, we're ready: we open the relief valve – which in orbit would let air escape to space – and the technicians seated outside turn on the chamber's vacuum pump.

When the pressure in the chamber drops below 0.4 bars, I can hear the usual creaking of the suit as it inflates. The space around my body expands, and the soft membranes against my legs and arms become stiff walls. These sensations are familiar, but I'm well aware that the pressure outside the suit is actually decreasing this time, as it would if I were on the Space Station and the air were seeping out into space. Eventually, the background noise from outside, which I hadn't even noticed, grows muffled before it completely disappears. We are acoustically isolated, apart from the technicians' voices coming through the audio system.

I try whistling. I round my lips and give it my all, but I manage only to let out a hiss and the occasional wheeze. Through the headphones I can hear Alex trying to do the same – and failing. Whistling is easy – but not when the atmosphere in the Orlan suit is less than half the pressure at sea level, too rarefied for your lips to produce the sound. Even our voices have changed, and our hoarseness is accompanied by a strange sensation in our airways: not unpleasant or alarming, just unusual.

I'm pleased with the muffled sounds, happy with the change in my voice and my inability to whistle. I'm happy about all these little signs because they remind me how unique this experience is. We are all surrounded by vacuum at all times, beyond the thin atmospheric layer that contains all life on Earth. But probably no other human being in the world is surrounded by vacuum at this exact moment in the same way we are. Our training as astronauts must transform the exceptional into the ordinary, rendering familiar experiences that are anything but. Every so often, though, it's good to take a moment to savour the extraordinary.

—

Russian EVAs in the Orlan suit are usually reserved for cosmonauts, in the same way that only Americans, Europeans, Canadians and Japanese are trained to use the EMU. Alex and I owed our unusual opportunity to a Russian module called MLM, or Multifunctional Laboratory Module. ERA, or European Robotic Arm, which was developed and built by ESA, was scheduled to launch with the MLM.

Per the agreements, once the MLM was launched, a European astronaut would participate in an EVA during the final installation of ERA. This was planned during Alex's stay on board, and he was therefore the designated astronaut. Since one never plans anything in space without considering potential delays, I would be trained as a back-up for this EVA. In truth, the MLM had had a rather troubled history, to the extent that its launch had been put back by several years. There was therefore every possibility that this fateful EVA would be postponed from Alex's expedition to mine. The potential for it to slip beyond that, however, was equally high.

Still, both of us were really pleased to have this opportunity. No matter what happened with the launch of MLM, if we were offered the chance to broaden our skillset as astronauts with such unique training, we certainly wouldn't complain – far from it. And for my part, I was glad for the extra hours underwater with a pressurized suit, even if the Orlan was very different from the EMU. For an astronaut, hours in the water are like flight hours for a pilot or, I can imagine, hours at the piano for a pianist, time on the slopes for a skier. There are certain things you learn only by brute repetition and practice.

It was a nice time to be in Star City, because a really fun and close-knit group of astronauts was training there. Besides the usual busy social life in Cottage 3, we often found a way to do things together at the weekend. One Sunday afternoon, for example, we went go-karting in Moscow, proving beyond reasonable doubt that no matter how well we controlled it most of the time, our competitive spirit was alive and well.

About that time, NASA astronaut Kevin Ford was preparing to leave for space and on one of his last evenings in Star City there was a video conference with the Space Station, which offered a chance for a final exchange with the on-board crew before his launch. After the official bit, Suni, then commander of the ISS, asked Kevin to invite all of us in to say hello. With their image projected on the large screen, it was almost as if the ISS crew were raising a glass with us at Shep's Bar. But we were the ones with the glasses – they were sucking from bags of soluble orange juice.

A few days later, we gathered again in Shep's Bar for another video conference, this time with Houston. We were eager to hear an important announcement: the name of the person who would be carrying out, a little more than two years later, a unique mission lasting a whole year, double the usual duration. Terry and I were particularly interested since we'd be welcoming this NASA astronaut aboard a few months into our mission. We discovered that it was Scott, whom I'd met on various occasions by then, always appreciating his frank, friendly and generous ways. I was sure he'd make a great crewmate. We still didn't know anything about our Soyuz commander, other than that it wouldn't be Sergei Zaletin, contrary to what they'd announced some months earlier.

While I was waiting to learn who our commander would be so I could train with him on the Soyuz, I dodged the Russian winter by heading for the mild Houston temperatures. The trip template meteorology was in my favour this year. It was a fairly brief trip, but the programme was extremely intense: along with some lighter courses, I was scheduled for three NBL runs over a span of just twelve days. I'd have to prepare for each of them more meticulously than usual, if possible, since this was about filling in training gaps that had shown up during the summer, after which I'd repeat the exam. Up to now, I'd had different instructors for every run, but now I was officially handed over to the team assigned to my crew. Faruq, who was responsible for underwater training, and Alex, who dealt with our EVA preparation overall, including suits and the Airlock, would accompany Terry and me for the coming two years. Being coached continuously by the same people proved extremely useful, and the extra runs greatly improved my awareness in the suit as well as my confidence. This time the exam went well, and the very next day I went back to Europe.

On 30 November, I got a message from Terry in Houston saying simply: 'Two years from today!'

14.

Star City, 11 April 2013

The large circular room is flooded with light shining through the tall windows, which look over the back of the building towards an area of untamed vegetation. Across from that area loom several mock-ups currently out of action. I linger here for a short time, leaning against the metal handrail that runs around the edge of the pool. I'm always fascinated to watch the platform as it's slowly lowered into the water, to the sound of large metal chains unrolling. As the mock-ups of the Russian modules on the Space Station go down, images of the parts that are already submerged are distorted by the refraction and blurred by water ripples. They look like they're coming from afar, and in a sense they are: below the surface, space begins.

A circular pool 23 metres in diameter, the hydrolaboratory is the Russian equivalent of Houston's NBL. Rather ingeniously, the mock-ups here do not sit at the bottom of the pool but rest on a mobile platform that can be lifted out of the water in just a few minutes. In Houston, I used to scuba-dive before a run in the suit, to explore the area I'd be moving and working in. Here, though, it's possible to step on the platform and explore on foot, a rather surreal collision between an everyday activity – such as walking – and this detailed, labyrinthine construction that evokes the outside of the Space Station. Alex and I had anyway the chance to go down to scuba-dive a few days ago, but this morning we've come to the platform for one last look at the translation path we'll follow along the Russian Service Module today, once we've exited the airlock. It's the first of three runs, and if we complete them successfully, we'll obtain a basic qualification for spacewalks in the Orlan.

Together we head for the crew changing room, furnished in the typical Star City style, so different from the colder, down-to-business character of similar European or American facilities. There's a long wooden table in the centre, a large squishy sofa decorated with a pattern of leaves in faded colours and flowery throw rugs on the floor near the two dressing areas delineated by wobbly screens. Three nurses in white aprons are in charge of operations in this room. Warm yet decisive, they help us position the medical belt with its heart rate and respiratory rate sensors. They then instruct us to get into the underwear and the blue cooling undergarment.

After a quick medical check-up, we go back to the pool and start for our suits, which are being seen to by technicians in khaki trousers and jackets. Some of them look quite old, and I suspect they've been working at the hydrolaboratory since it was built in the 1980s. A technician reminds us that the water version of the Orlan has a safety feature: the hinged backpack, which in space houses life support systems such as the oxygen supplies, will instead contain emergency air tanks. If the feed coming through the umbilical from the surface gets interrupted, the suit's occupant can easily switch to the tanks by flipping a lever on the chest. Just beside that is the large knob of the pressure regulator. I could rotate it myself, but I let the technician do it. It's pointless to start tiring my hands now – I'll have plenty of opportunity to expend my energy later. The Orlan works in fact at an overpressure above that of the EMU, so it's stiffer, and every movement demands an even greater effort. The needle on the pressure gauge stops at 0.4 bars. Of course, a higher pressure also carries with it a number of advantages: to begin with, a wider safety margin in case of a leak, and a very short pre-breathe protocol of only thirty minutes before an actual EVA in space.

As always, the Orlan goes on very quickly. Around fifteen minutes from our arrival at the pool deck we're already in the water, and handed over to the divers, who take us to the bottom to work on our weigh-out. Our instructor, Valery, watches us from his post behind a large window in the wall of the pool. Other, smaller windows allow the occasional visitor to watch the activity, like tourists visiting an aquarium.

When the weigh-out is complete, the divers take us to the airlock, and from that moment onwards, we're in space, completely reliant on our hands to move us along the structure. Today, the first of the so-called typical operations we'll perform are opening the hatch and exiting the airlock, using all the safety protocols. After this, we'll translate along the Service Module, each carrying our own bag, and then a larger load we'll have to manage by helping each other. We'll be working with small tools and electric connectors and we'll practise operating Strela, a telescopic arm we'll learn to extend and orientate. All this should take no more than four hours.

Even more than I did with the EMU suit, I'm trying to plan every move accurately before performing it, and I strive to make slow, measured, regular movements, a little like the small steps you take when you're hiking in the mountains with a heavy backpack. The protocol for translating along the Station here is more laborious than the NASA protocol. We don't have a coiled safety tether, but there are two tethers attached to the suit, which have to be released one at a time – never together – in a sort of 'via ferrata' choreography. One hand detaches a tether and puts it down somewhere along the route, while the other hand holds on to something: it absolutely must not be the handrail to which the second tether is still attached, since it's not impossible that the handrail itself might come loose. At a certain point, I feel like I've ended up in a blind alley: I can't see enough handholds within reach that would allow me to continue, given all the rules. I pause for a moment, reminding myself not to give in to frustration, and then I look again. If nothing else, looking around in an Orlan is a joy. The helmet is roomy, and it has a wide field of view, with the great luxury of a little window on the top that allows you to see above your own head. In the end, the solution, as so often the case, is to reorientate myself in order to make the most of my limited arm span. I don't know if I'll ever perform an EVA in space, but training underwater has certainly taught me one thing: when it seems there's no way out – stop, breathe and change your perspective. A platitude, perhaps, yet one I feel I've learned in my flesh instead of my mind.

—

I'd looked forward to the training in the Orlan suit with curiosity, yet also a certain amount of apprehension – not only because it is stiffer than the EMU, but also because the Orlan comes in only one size. At 5 feet and 5 inches, I was the shortest acceptable height. There are various adjustment points, but they're all the same type: you can shorten the arms, legs and pelvis by pulling various cords that pleat the fabric, a little like you do with Roman blinds. The more you shorten it, the more the material bunches up. Not so bad for your legs, which aren't much use in space, but the thicker fabric around your elbows can be a hindrance if you want to bend your arm. As for the gloves, they come in three sizes, differing only in finger length. And you can't insert any sort of padding while you're training in the pool, as you can with the EMU.

The Orlan's intrinsic simplicity and sturdiness mean that while it is probably safer, it is definitely more limited in its capabilities than the EMU. On the other hand, the long, complex EVAs that were required during the assembly of the non-Russian part of the ISS or for its occasional repair are not envisioned for the Russian segment. The assembly of the Russian modules has demanded much less manual labour outside the Station, and almost all the critical elements can be found inside, in the pressurized volume, where they can be repaired without resorting to an EVA.

Despite the difficulties, I was pleased with my Orlan training: I managed to carry out all the required tasks, both underwater and in the vacuum chamber. My limited experience – around twelve runs in the EMU – had been helpful, even though it was acquired in a different environment. The reverse also proved to be true: I noticed in the coming months that, thanks to my training in the Orlan, I'd gained a new awareness which was really helpful when I came to train with the EMU in Houston. The Russian instructors placed great emphasis on managing physical effort, even letting the doctor intervene when necessary to make the astronauts take a break if their heart rate rose too high. It was important to demonstrate that you knew how to work at your own speed, avoiding bouts of exhaustion. I remembered something Suni had said to me during our first run together at the NBL: 'If you're working hard, you're working too hard.' I'd had to get to know the Orlan to really understand her.

I felt well rested and recharged when it came time for the course in the hydrolaboratory thanks to a light training load during the winter. This was totally unheard of, with only two years to go before the launch. At this point you're normally thrown into a marathon of training with no breaks, or almost none. As an ESA reserve astronaut, though, I'd set out on this marathon with a head start, chiefly because I'd already taken the theoretical course as Soyuz flight engineer. You could say I'd covered a good 10 kilometres before the starting signal. In my naivety, I'd dreamed of being able to profit from this advantage along the way, taking short rests now and then during the twenty-four months left before the launch. This illusion was quickly shattered. For practical reasons to do with planning and organization, I had to sync as quickly as possible with the rest of the crew members whose schedule was depicted in the little coloured boxes on the trip template.

That winter I was in Houston for only a couple of weeks following a quiet New Year with a few friends and a brief family visit to California. These were fairly light weeks – no NBL runs, but quite a varied programme of classes, most of them with Terry. One of them involved a simulation of one of the most-feared breakdowns: a critical bus failure, meaning the loss of one or more power channels from the ISS solar panels, a circumstance we were to simulate at regular intervals up to the launch. Depending on which channels are lost, the implications can be extremely serious – for example, one of the two pumps that circulate ammonia in the external cooling lines could shut down. Many vital pieces of equipment will immediately begin to heat up, setting off what's called a thermal clock, a race against time to reconfigure the Space Station before the machinery begins to shut down or is damaged due to overheating. Of course, the crew can usually rely on support from Mission Control, whose specialists are far more experienced with those complicated recovery procedures. But what if you lose radio contact, a likely situation in such circumstances? In that scenario, we have to execute dozens of pages in the rather intimidating Procedure 2.600. I find it amusing that it starts out with a simple lights check: if you suspect an electric bus failure, the first thing to do is to turn all of them on across the different modules in order to determine which power channel is affected, according to which lights stay off. I assume that the controllers on Earth have more sophisticated methods, but this was clearly judged to be astronaut-proof. Despite the prevailing image of us as particularly brilliant people, the specialists working with us are well aware that we have to know everything – but of that everything, in fact, we inevitably know very little.

When I got back to Europe, I made the most of a six-week break to take care of everything I'd left hanging during my last year and a half of world travel. I then resumed training with the experimental science racks in the European Columbus lab, and preparing for the certification exam at the 'operator' level. As an operator, I would be trained to respond to any malfunction serious enough to require timely intervention to prevent serious damage to the on-board equipment.

The time came to leave for Russia, and, as always, when I was back in Star City, it felt like I had never left. There was something of a change, however. For the first time, I was there not for generic training on the Soyuz, but as a crew member. The classes were the same, and so were the instructors, but there was a subtle difference, something I couldn't put my finger on yet was clearly noticeable in the way I was treated. I was no longer just Samantha, but Shkaplerov's flight engineer. That's right: in Star City, crew members are identified by the name of their commander, and ours had been named some time ago: Anton Shkaplerov. The same Anton who had casually toasted, in my house, to our flying to space together one day. I felt so fortunate. Not only was he an expert commander, recently returned from his first mission on the ISS, but he was cheerful, friendly and kind. His radio callsign was Astrey, the Russian name for the titan Astraeus. According to Greek myth, Astraeus' union with Eos gave birth to the stars and the wind, including the cold north wind, Boreas, the name of my class at the Air Force Academy. In keeping with custom, Astrey became the name of our crew.

We were also assigned a Soyuz instructor, Dima: young, affable and very conscientious. He managed our simulator training sessions, which were scheduled in four-hour blocks over the week and were always preceded by two hours of classroom instruction. At first, Anton and I were on our own, but we were soon joined by Terry from Houston. For the first time, the entire Astrey crew was training together to fly into space.

15.

Noginsky, 28 June 2013

Let's take a rest, Anton suggests, handing me a bottle of water. I want to drink the whole thing, but I take only a few sips. There's at least an hour to go before we can open the hatch of this old Soyuz descent module and jump in the water. We've finished the first task – removal of the Sokol spacesuit – and now, sprawled in our respective seats, we're enjoying a few minutes of doing nothing, trying to get our heartbeats and body temperatures back down to normal. Today, the road to space goes through here, a small lake in a Russian civil protection training centre not far from Star City. This is where Anton, Terry and I simulate a situation in which our descent module ends up in the water after an emergency re-entry. Outside, it's 35° C, and the summer sky is cloudless. Each of us has a sensor in our bodies; we swallowed it this morning, and it's now transmitting our core temperature to the NASA doctor overseeing this drill, since heatstroke is at the top of the risk matrix for this activity.

We try to avoid it, of course, while we practise the procedures for abandoning the descent module and preparing for a water recovery by the search and rescue teams. Despite our desire to get out as quickly as possible, we move unhurriedly, one at a time, helping one another and following the considered suggestions of Anton, who has already performed this drill several times. Lobster-red and soaked with sweat, we're beginning to forge strong bonds, to invest in mutual trust and shared experiences, all of which will help us to get through a long space mission together. When it comes down to it, it's the little things that build a crew: helping each other release hooks and zippers to minimize effort; supporting one another while we stretch across each other to wriggle out of our suits; ignoring the unintended jabs from elbows or knees; directing the air flow from our own ventilation hose towards a colleague as they struggle to get the suit neck ring over their head, a difficult operation even with both feet planted on the floor. Almost unconsciously, we notice a kindness that relieves discomfort, a joke that smooths over a situation or an observation that confirms clarity and alertness. No one gets irritable or nervous. Maybe we take it for granted; we'd definitely notice if it weren't the case.

After a brief rest, we start gathering the clothes we'll need. First, we put on the so-called flight suit – lightweight blue dungarees – followed by a thick winter suit and then we prepare to put on the Forel, a somewhat old-fashioned orange dry suit that's quite tough to get into. From this point on, we must be decisive and act quickly, because core temperature rises fast in the Forel. We help each other get into the suits and then tie flotation belts around our waists; we'll inflate when we jump into the water. A final quick check – and we're all ready to leave and get some relief from this heat.

Now our exercise starts to take on a much faster pace. Permission to exit comes over the radio, and Anton opens the hatch. We climb up awkwardly to reach the rim and throw the survival pack, which is tied to us with a cord, in the water – one pack for each of us. I'm the second to exit. Leaning forwards at the hip to hold on to the rim of the hatch, I extend my legs and then let myself fall backwards, taking care not to rock the descent module; our instructors impressed us with the wisdom of this during our days of prep. After an hour and a half in that exhausting heat, contact with the water is a pleasure, instantly refreshing.

In a genuine emergency situation the pleasure would no doubt be brief, especially considering that the Soyuz doesn't have any life rafts like those usually attached to the ejection seats on combat aircraft. In water, even at non-freezing temperatures, hypothermia is never far off. We don't need to worry about that today, however, since the drill will be brief. We only need to demonstrate our proficiency in using the survival kit – in particular, the signalling tools, from the radio to the emergency flares and even the classic mirror, which is simple and reliable and allows you to attract the attention of the rescue teams when you point the reflection of the Sun in their direction. We carry out all the procedures without much trouble, well aware that in an actual emergency the conditions could be much more difficult: we might find ourselves in a roiling sea, it could be night, or we could be put to the test not only by re-entry through the atmosphere but also by bouts of nausea inside the capsule. Water survival commands respect. Every astronaut would always prefer to land on solid ground or, given a water landing, would rather stay inside the descent vehicle, circumstances permitting.

Yet what if it's necessary to make a rushed exit, with no time for the luxury of getting into warm, waterproof clothing? What if, say, you realize that you're sinking? Yesterday, to prepare for the possibility of a rapid evacuation we carried out the 'short drill', in which we abandoned the descent module in a few minutes, having fastened only the flotation belt over the Sokol suit. It's true that the descent module should float, even without the inflatable orange safety ring that surrounds our drill capsule today – but there's no guarantee. Just think of the large niches that hosted the parachute and are left empty after its deployment; they could take in water. To prevent that, inflatable bags should fill up the empty space automatically, but all automatic mechanisms are subject to failure. Or think of Soyuz 23, which ended up in Kazakhstan's Lake Tengiz in 1976 – as of today, the only emergency water landing yet. Because of adverse weather conditions, recovery was not possible for several hours, and the soaking parachute began dragging the re-entry vehicle to the bottom of the lake. It may be for this reason that today, procedures see us releasing both parachute cords immediately in case of a water landing. In spaceflight, as in aviation, procedures are often the legacy of colleagues you've never met, but who lost their lives in a plane or a space vehicle. Luckily, the crew of Soyuz 23 were rescued alive – somewhat to the surprise of the search and rescue team themselves.

The drill over, we quickly change clothes and take a moment to rehydrate before joining our instructors for the debriefing. We settle into some chairs set out on the grass beside the lake and talk first about procedural and technical aspects of the exercise. Then a young psychologist speaks. Smiling broadly, she observes that I dropped into the water in a particularly feminine way. For a long time now I've been resigned to not understanding everything psychologists have to say, so I decide it's a compliment and give it no further thought. The head instructor concludes the debriefing process – and with it, our week of training in the water – with the usual hope that we'll never have to put into practice what we've learned over the past few days.

———

It's a wish that applies to almost all our training on the Space Station. We spend a large part of our time preparing for situations we hope will never arise, from fire on board the ISS to a depressurization on the Soyuz.

If one could rule out beforehand any sort of failure or emergency, the training would only take a few months. And that's how it actually was for the 'spaceflight participants', a few very wealthy people who took advantage of the occasionally empty seats on the Soyuz in the days when the Space Shuttle was still in service as an alternative means of transportation for the ISS crew, and paid extravagant sums to realize their dream of going into space. They were all capable individuals, but in an emergency situation they'd have had to rely on the professional astronauts with their years of training behind them.

The irony is that you are often better prepared for a serious emergency than you are to carry out your daily work on the Space Station, which mostly consists of routine activities and somewhat trivial tasks, such as managing inventory or doing small maintenance jobs. Don't think for a moment, however, that the daily activities are completely overlooked during training. Actually, in Houston, we periodically get together with our crewmates and a large group of instructors for the so-called routine ops sims, simulations lasting for several hours when – blissfully – no alarms go off in the Space Station mock-ups and the most dreaded incident is having to change a full solid-waste container from the toilet. It was actually that spring, before my trip to Russia for the water-survival training, that this sort of simulation presented the first chance for Terry and me to train with Butch and Scott. Josh, our chief training officer and the one responsible for coordinating the work of all the Houston instructors, oversaw the simulation. He worked out a typical activity timeline, which included the collecting and packing of return cargo for a virtual departing Soyuz; taking air samples and surface swabs for periodic analyses; on-board computer maintenance and changing the canister with the brine left over from urine recycling. Everything was posted on a dedicated timeline viewer, which you can access in orbit on computers throughout the ISS. When you click on any given activity, you gain access to support material, such as procedures, execution notes with specific instructions and stowage notes. Despite the glamorous image of astronauts, and although we're trained to manage emergency situations or perform sensitive activities such as spacewalks, most of our work in orbit consists of this: consulting the procedure assigned to us at any given time, reading the execution notes, collecting the tools and other items listed in the stowage notes and following the procedure step by step. It's no surprise that a Belgian colleague used to say, with characteristic self-deprecation, that there's only one skill an astronaut needs to have: the ability to read.

Every joke contains a kernel of truth, but there are times when we do have to do a bit more. To give you an example, it just so happened that the ISS crew at that time noticed a stream of white particles like snowflakes seemingly flowing from a location on the far end of the truss. A few hours later, specialists from Houston's Mission Control Center confirmed that cooling ducts were leaking ammonia from one of the cooling circuits on the power channels, with the result that one of the eight solar panels would soon be out of action. It was hard to tell exactly where the leak was located, but they suspected it was the pump that keeps the ammonia in circulation.

There was only one way to find out, and that was by sending two astronauts to change the pump and to ascertain whether that had stopped the leak. There was limited time available to carry out this spacewalk: the Canadian Chris Hadfield and the American Tom Marshburn were scheduled to return to Earth the following Monday. There would then be only one NASA astronaut on board, Chris Cassidy, and although he was an exceptionally skilled spacewalker, it was unthinkable for him to go out alone, as it would be to ask the Russian crew to carry out this sort of repair in an Orlan suit. New astronauts trained in the EMU wouldn't arrive for another two weeks, enough time for all the ammonia in the affected circuit to disperse into space. Not only would it be necessary to replace it, but without the stream of white flakes, it would be impossible to determine where the leak was located. That Thursday, meetings and analyses went on into the evening and beyond, and on the Space Station, five hours ahead of Houston, the crew went to bed without knowing the outcome. When they woke up Friday morning, they found a message from Mission Control: 'Welcome to EVA preparation day.' The next day, Chris Cassidy and Tom would carry out a contingency EVA. Chris Hadfield would manage the operations pre- and post-EVA in the Airlock, with a specific twist: he had to be prepared to initiate decontamination procedures should the two spacewalkers be struck by ammonia spray.

This was an exceptional occurrence: never before had a spacewalk been scheduled with only one day of preparation. In Houston, the EVA community was working feverishly. I met some exhausted people who hadn't been to bed the night before and probably wouldn't be able to go home until the next afternoon, and by coincidence Terry and I also found ourselves contributing to the effort to send Chris and Tom out the door with the best chances for success.

We were actually scheduled for a training run with the EMU that very day. Given the circumstances, however, the NBL pool would instead be used to test procedures being developed for the unexpected spacewalk. It wasn't possible to replace Terry or me with EVA veterans since our suits were all ready and waiting for us, assembled to our measurements. That's how, despite my limited experience, I found myself underwater, confirming procedures and operations sequences, work envelopes and bag and tool configurations so that the specialists could prepare in real-time an updated briefing package to send to orbit before the astronauts went to sleep. When we emerged from the pool in the middle of the afternoon, it was already night on the ISS.

After six hours of hard work in the water, unless I am completely disappointed with my performance, I always feel quietly pleased, with that sense of deep satisfaction you only get after doing something strenuous and difficult. On that particular day, along with the usual sense of fulfilment there was a smidgen of pride for having contributed in a small way to preparations for Chris and Tom's spacewalk. I had never been this close to critical operations in space. The next day, I happily got up early to follow the spacewalk from Mission Control. I was allowed to listen in on the private channels of communication between the specialists and this furthered my understanding of how decisions are taken in real time. The EVA was successful, and the pump replacement halted the ammonia leak. The power channel was restored, and, as expected, Tom and Chris Hadfield returned to Earth the following Monday, with the world going crazy over Chris's now famous cover version of David Bowie's 'Space Oddity', actually performed in space. For Chris Cassidy, however, the experience in orbit would continue. There were several more surprises in store for him.

16.

Star City, 24 July 2013

People are bustling around me, some in khaki uniforms, others
in white aprons. They're tightening straps, attaching cables, positioning sensors, adjusting things for my height so that I'm comfortable and well supported. I'm in the position I'll be in on the
Soyuz – back parallel to the ground, legs bent – though this seat is
much more comfortable. A heavy metal ring, or a sort of mechanical halo, is on a level with my face. Irina Viktorovna, one of
the doctors in Star City, comes over and explains that the tests for
visual acuity will employ a series of small lights situated on the
ring, 80 cm in front of my eyes. Energetic and determined, Irina
exudes competence and practicality. During a brief visit a short
time ago, she had me lie down on a bed while she put her hands
on my sternum, leaning into me with all her weight so I could
practise the correct breathing technique: it's crucial for protecting your heart and ensuring that it continues to function normally while your ribcage is subject to unusual pressure. I'm about to
be spun around at 8 Gs, in the world's largest centrifuge.

The preparation is completed and the technicians push my
seat into the cabin, which is hinged on one end of the imposing 18
metre mechanical arm. There's a metallic sound as the seat locks
into position, and then I hear the hatch close behind me. The
heavy access doors halfway up in the tall walls of the large, circular hall will also surely be closing now. I've watched the centrifuge spin a few times – you can see it through the large windows
of the control room above, revolving in all its majesty. All told,
it's quite a simple machine, but powerful in its ability to combine

mass and velocity. Some people like precision machinery, say, the workings of a wristwatch, but I have always been fascinated by colossal machines.

Unlike the Soyuz descent module, this cabin is really spacious and somewhat bare, dimly lit by a light that falls mainly on my face. I know that I'm being carefully observed by Irina Viktorovna and the rest of the medical staff. They can see my face in the video images, and the medical belt I'm wearing allows them to monitor my heart and my breathing. I've also got a cuff on my forearm that will periodically inflate to take my blood pressure.

For several minutes there's nothing but silence. Then a voice comes over the speakers to say that the cabin is moving into departure position. The centrifuge starts rotating slowly for the first warm-up exercise at 4 Gs. Increments of 0.1 Gs per second are planned, and the arm accelerates its rotation accordingly. Irina Viktorovna announces that we are passing the 2 G mark, and I begin to feel pressure on my chest. At 3 Gs, I am still able to breathe normally, but I start to use the breathing technique I've learned: I contract my chest and neck muscles, thus stiffening my ribcage and airways, and I use my diaphragm to replace some of the air in my lungs. It's not easy to coordinate these two actions, and I've never had much practice with the technique, though it's a little like the abdominal breathing exercises I've practised in yoga lessons over the years. I've often heard veteran astronauts say that in this job everything you've ever learned proves to be useful at some point or another – and it's true.

The centrifuge stops for a minute at 4 Gs, then begins to slow down. The warm-up is over, and it didn't bother me apart from a slight pain in my throat. Next stage: 8 Gs.

Up to now, it's been a bit of a déjà vu. Last week I experienced in the centrifuge the acceleration profiles for a nominal launch and re-entry in the Soyuz, and during the ascent the peak is in fact around 4 Gs. It occurs at the end of the first-stage burn, when the full thrust from the four side boosters is still available, but the rocket has now become lighter as it has shed the weight of all the propellant it's already used. That thrust, although produced in a very different way to what occurs in the centrifuge, exerts the same pressure on the chest. Also during the re-entry through the atmosphere, with two typical peaks of 4.3 Gs, the deceleration is felt by the crew in the same direction: chest to back. It's no coincidence: this is the direction in which G-loads are withstood most easily and with the least risk. The acceleration from the head towards the feet, which you get in aeroplanes, and is particularly intense in acrobatic or combat manoeuvres, is much more insidious since it can cause blood to flow away from the brain, leading to a temporary loss of vision or, in the most extreme cases, loss of consciousness, or G-LOC. To mitigate this risk, pilots are trained to use a special breathing technique and contract their leg, buttock and abdominal muscles to prevent the flow of blood towards the lower parts of the body. The G-suit, which inflates automatically as necessary, contributes in turn by compressing the blood vessels in the lower limbs. Despite precautions, G-LOC does happen, sometimes with fatal consequences. So it made sense that, given the choice, designers of space vehicles decided from the beginning to position the seats in such a way that the effects of acceleration would be directed towards the back.

Given this favourable design, and never having found it difficult to withstand Gs on an aeroplane, I've never been very worried about the centrifuge. And indeed, last week's spinning at 4 Gs went without a hitch. I am, however, a little concerned about 8 Gs, though I'd never admit it and I certainly hope no one notices. First of all, I expect that at some point, discomfort will become pain. More than anything else, though, I find it worrying to have a plot of my heart function transmitted in real time to the medical console. It wouldn't be the first time that little anomalies in heart function were observed in the centrifuge, and it's well known how attentive the Russian doctors are, occasionally picking up problems on the plot that wouldn't concern their Western counterparts. Like any astronaut, the last thing I want is to be an object of discussion for the medical board.

The centrifuge starts up again, and this second spinning,

somewhere between 4 Gs and 8 Gs, starts to feel rather uncomfortable, causing an unpleasant ache in my lower sternum. Unpleasant but bearable. The 8 Gs portion will only last for thirty seconds, after all. I know it's not the case, yet it feels like my lips have stretched all the way back to my ears. I feel tears running down the left side of my face and I know those are real. They told me this might happen: at 8 Gs the eyeball becomes slightly distorted. It's a temporary effect and nothing to worry about, but it explains the purpose of the visual acuity test I'm about to take. Would I be able to read the instruments and displays in these conditions? Irina Viktorovna spells out the instructions through the speakers. The little lights on the metal ring above my face go on in rapid succession; sometimes they're placed centrally and sometimes at the margins of my peripheral vision. As soon as I see a light go on, I have to turn it off as quickly as I can by pressing a button I hold in my hand. It's a simple test of my field of view, and it's followed by another standard test in which I have to determine on which side there's an opening in a series of circles. With some effort, I manage to make out the smallest one on my far left. I don't say anything, however. Irina Viktorovna will ask me for my response later. Though I'm somewhat tempted to try, her instructions are clear: you don't speak at 8 Gs.

—

It's embedded in the collective imagination that astronauts regularly practise in the centrifuge and are able to withstand the tremendous acceleration to which they're subjected thanks to their exceptional physiques and tough training – essential preparation for enduring the rigours of spaceflight. The idea took root in the pioneering era of space travel and is perpetuated by countless documentaries showing the centrifuge making an inevitable theatrical appearance to the sound of pompous narration and dramatic music. Personally, I'm more afraid of the cold. Any healthy person without psychological issues and able to follow simple instructions can withstand 4 Gs; with patience and some tolerance of physical discomfort, even the 8-G exercise. The latter isn't intended as a test of an astronaut's stamina; it's meant to familiarize them with conditions that will be experienced in case of a ballistic re-entry, when the load factor may reach 8 Gs or even exceed that by a considerable amount. Ballistic means unguided: the capsule flies through the atmosphere passively like a rock, affected only by the forces of gravity and aerodynamics, except that it also rotates about its own axis at $13°$ per second. Unlike a rock, however, the descent module is shaped like a bell, and its mass is distributed so that it will orient itself with its chunky base, protected by a heat shield, in the direction of motion. A ballistic re-entry is definitely uncomfortable, but it's safe. The Soyuz has brought crew home several times after serious failures thanks to this emergency descent mode, which is solid and reliable.

Many types of malfunction can lead to an unguided re-entry, even in the later stages of descent. For this reason, there are always two search and rescue teams waiting for a crew returning to Earth: the first at the nominal landing site, and the second at the ballistic one. If, however, the re-entry is unexpected – caused, for example, by an emergency on the Space Station – a ballistic re-entry is the only option possible, because the data necessary for a guided descent will not be available to the computer. At that point, you have at least to switch on the engine at the right moment, and that timing is indicated on a special chart called Form 14; this is sent by the Russian Mission Control Centre TsUP every day and is printed and put up in the Soyuz by the commander as the first action of the day after waking. Switching on the engine for a re-entry according to Form 14 means that after a difficult ballistic descent, landing will take place at the most hospitable site available on the ground track of that particular orbit – instead of in the middle of the Pacific, for example, or on top of a mountain in the Himalayas. Either way, there won't be any rescue teams waiting.

Getting back to Earth alive is dependent on many factors, and astronauts aren't able to intervene in all of them. Only the descent module, for example, can withstand the heat of the re-entry. Before the Soyuz enters into contact with the atmosphere, the descent module separation must be completed successfully. If the pyrotechnic devices that separate it from the orbital and service modules don't ignite at the right time, the crew can send the command manually. If the charges still don't ignite, the entire vehicle will burn up in the atmosphere, and the crew won't be able to do a thing. At 10 kilometres up, the parachute opens automatically. If it doesn't open, there's a reserve parachute. If that one doesn't open either, the descent module will be destroyed on impact with the ground and, once again, there's nothing the crew can do about it.

However, there are other situations in which you can and must intervene. If the computer fails, the crew can guide the re-entry using an auxiliary computer, or indeed manually. And if the main engine doesn't ignite or shuts off too soon, the crew can use the small attitude control thrusters to provide the required braking burn at the right moment for the planned duration and in the correct attitude. This is the key for re-entry to Earth: slowing down just enough but not too much, so that a little later, the Soyuz slips through the upper atmosphere at a precise angle, steep enough to keep it from bouncing away, shallow enough to keep it from hurling itself destructively towards the Earth's surface. Slowing down at the right moment enables it to land at the designated spot in Kazakhstan, where the rescue team is waiting.

Most of Soyuz training concerns just this: in the simulators, we learn how to perform a re-entry successfully when confronted with unlikely combinations of failures. As far as possible, we try to stick with the automatic, guided re-entry with its peaks of only 4.3 Gs, but we also learn how to promptly recognize the need for a ballistic return. Between these two extremes there are intermediate types of failures in which the computer is unable to guide the re-entry and you can assume manual control of the vehicle and still try to fly along a descent trajectory that's comfortable in terms of the G-load which the crew will experience, and precise in terms of its proximity to the expected landing point. There has never been a manual re-entry in the history of the Soyuz, but every commander and every flight engineer is trained to perform one.

There's a specific simulator capable of generating the varying initial conditions of being early or late when making initial contact with the atmosphere. Being early means that you have to fly along a shallower trajectory to prevent the parachute from opening too soon so that you land short of your intended landing site. Conversely, a delay requires a steeper trajectory and constitutes a much more dangerous situation, because if it's not perfectly controlled, it can quickly bring about an elevated G-load. The Soyuz doesn't have wings or control surfaces like a plane's ailerons or its rudder. It can, however, generate a small amount of lift, which it can regulate thanks to a carefully chosen offset of the descent module centre of mass from its axis of symmetry, and a set of thrusters that can make it roll: rolling the capsule in one direction decreases the lift, the descent module follows a steeper trajectory and the landing site moves closer. Rolling in the opposite direction will achieve the opposite effect.

In a nominal situation, the computer regulates the roll, and the crew only monitors the descent on a dedicated page on the command and control display, which shows the calculated trajectory alongside the actual one as the re-entry progresses. The computer is very precise, much more so than a human could be. Just think: in an exam, you receive top marks when you complete the descent – and so the parachute opens – within 10 kilometres of the planned location. But the maximum acceleration achieved should also be considered; it should never exceed a limit of between 4 Gs and 6 Gs depending on initial conditions. Quite apart from your exam marks, you would literally feel the consequences of an excessive G-load first-hand in your body. The re-entry through the atmosphere may be simulated, but the Gs are real, since the exam takes place in the centrifuge.

After the two familiarization runs, I would sometimes spin at the end of the long mechanical arm, no longer a passenger, but actually at the controls, by means of the re-entry simulation. In front of me would be the Soyuz control panel, and I'd hold the descent control device, an old-fashioned-looking gizmo connected by a cable to the on-board systems. Two large handles at the side allow you to hold it comfortably and securely, even if you have to grip it wearing Sokol gloves, which may actually be pressurized due to an ongoing emergency. It's easy to push the two control buttons with your thumbs:

one rolls the descent module in one direction; the other makes it roll in the opposite direction. Two buttons. That's it.

It wasn't instantly clear to me how such elementary commands could be translated into the correct management of the trajectory. After the first simulator sessions, I was quite puzzled: the Soyuz was not only rather sluggish in its response to commands, but also quite unpredictable in its behaviour, due to the unequal density of the different atmospheric layers and the varieties of centres of mass that were simulated. During a supper at Cottage 3, after one of the first drills, I made my colleagues laugh when I jokingly said that in some simulations flying the Soyuz was like flying a glider, in others a clothes iron. Their laugh was one of kindness and understanding. All the astronauts who'd come through training before me had gone through the same experience. They gave me an unofficial cheat sheet, handed down from one crew to another and adapted by each according to individual preferences. It wasn't the solution to all problems, but if nothing else, it allowed you to handle the beginning of the descent correctly without thinking about it too much. What do you do if your contact with the atmosphere is delayed by thirty seconds, for example? Immediately initiate a 45° rotation to the left in order to start the steepest descent possible; but after you pass 3 Gs, you promptly rotate 45° to the right, in order not to exceed 5 Gs, the maximum allowed. It was a safe starting manoeuvre, allowing time for you to observe the vehicle's behaviour: from that point on, it was an art.

Whenever I found myself back in training in Star City, I had two or three simulator sessions a week with Dima, who patiently loaded one scenario after another. Repetition is the mother of learning, as the Russians love to say. In fact, things did slowly improve, and I was relieved to discover one day that I wasn't getting it wrong any more. I'd joined the ranks of those who'd overcome the initial confusion, and I could now begin passing on my experience to the astronauts coming up after me in training, just as Reid and the other flight engineers from earlier crews had done for me.

Luca, too, had shared his techniques for coping with the re-entry simulator. We hadn't seen each other much over the last three years, because Luca and his family had moved to Houston right after his assignment. Every now and then we'd run into each other in some part of the world or other, when the little boxes on our trip templates showed the same colour. I'm the kind of person who needs regular doses of silence and solitude, but Luca is extremely extroverted, always eager to share his stories and experiences. Even just a short time before we said goodbye in Star City ahead of his departure for Baikonur, he actually took the time to explain techniques he'd found useful, not only in the descent simulator, but also in manual docking.

At the end of May, I followed Luca's launch from ASI headquarters in Rome, commenting from the stage on the images transmitted from Baikonur. There was a large crowd, including his family and friends. We all burst into applause, relieved when shots from inside the Soyuz showed floating objects – Luca, Karen and Fëdor were in orbit. You don't indulge in your own, heartfelt emotions in the midst of a crowd, but later, when most of the guests and journalists had gone away, knowing that Luca was in his Soyuz moving towards the Space Station really affected me. The first of the Shenanigans was in space. It was four years since the selection, a short time compared to the average time an astronaut waits for a first flight. It seemed like yesterday that we'd crossed the threshold of EAC together in Cologne, but we'd learned so much since then, had had so many experiences and met so many people – Luca, the other Shenanigans and I. And now Luca was in space, something both expected and extraordinary. Three years before, I wished I had been assigned to that flight myself. But now I felt almost grateful not to be in his shoes. Luca was ready, without a doubt, but it seemed to me that it had all come too quickly for him.

17.

Moscow, 1 October 2013

These scales have been in use since 1961, says the metal label. In their time, they've weighed Gagarin and Tereshkova. By now I'm used to it: last year, I was measured here, at the headquarters of the suit manufacturer Zvezda, with an equally glorious stadiometer bearing labels with the height of those two pioneers. It seems Gagarin was only a couple of centimetres taller than I am, and Tereshkova a tiny bit shorter. I'll never know what they weighed, however; I've searched in vain for similar labels on the scales.

After the weigh-in, it's time for a series of anthropometric measurements. An elderly man with a kindly expression holds the ruler, while another man, somewhat hunched, looks on, scowling, and an earnest lady carefully registers my measurements in a thick notebook. I wonder if Gagarin's and Tereshkova's measurements are also there, in the first pages. But no, they can't be. The first notebook must have been filled up some time ago. Nevertheless . . . I'll take a peep.

Today is a momentous day. At little more than a year from the launch, it's now time to take the measurements for my Sokol spacesuit. Seventy of them, to be precise. And when they have them, people I will never meet will get to work somewhere in this building on the outskirts of Moscow, manufacturing the suit I'll wear when I fly into space. We're about to make a plaster cast of my body, too, and another group of people will use it to craft my seat liner. The scene looks resistant to any change whatsoever, as if it were suspended in time. In the middle of a room with custard-coloured tiles sits an object which at first sight looks to be a small but sturdy metal bath. Dressed in a one-piece cotton undergarment with long sleeves and a hood, I walk up to a knot of people in white aprons swarming around the tub and take my place in it, assuming the now familiar foetal position. Now preparations for pouring the plaster begin: they ask me to extend my arms upwards while the kindly looking man shields my face with his hands, and the earnest lady with a stocky build puts one hand on my abdomen and another on my chest, preparing to press down as much as necessary to stop me from floating. With this bizarre tangle of hands in place, a couple of assistants start pouring plaster from two buckets encrusted with the remains of who knows how many similar sessions. The mould is made in two sections: first the top part of my body down to my hips, and then my pelvis and bottom. When the casting is finished, the only person standing next to me is the lady, still resolutely pressing me towards the bottom of the tub. After the plaster has hardened, I'll be able to stand up and have a snack of instant coffee and biscuits while two technicians finesse the mould with a putty knife, eliminating all the excess material. Once they're happy with it, I'll be asked to take up my post again so I can point out any place where the mould should be improved. It's important to have uniform contact along the entire spine, especially at the neck, in order to distribute the load evenly. That will minimize the risk of trauma at the moment of impact with the ground at landing. I point out several pressure points and the technicians go back to work making the relevant adjustments. I wonder how they decide how much margin to leave for the suit and for the lengthening of your spine in space. Experience and intuition, I guess. A couple of centimetres for someone small like me, three or four for a larger astronaut? There don't seem to be any precise rules, but I'm not sure I understand this process very well. They don't speak much, those technicians – mostly amongst themselves.

I'm not worried. After all, the Soyuz has been flying for decades, and, as far as I know, not a single astronaut has suffered serious trauma. But at the same time, there's no denying the importance of making a good mould. I've heard that the impact with the ground is fairly brutal, though I suspect that comparing it with a car accident reveals something of a taste for exaggeration. At least according to the documentation I've been supplied with, the speed at contact must be around 5 kilometres per hour, assuming that the retrorockets are functioning correctly. The Russians call them 'soft-landing engines' – ironically, I hope. They ignite in a momentary blaze immediately before landing, activated by a signal from the radar-altimeter at about 1 metre above ground. If they didn't work, impact would come at 30 kilometres per hour, strong enough to crush the shock absorber in the seat, which would then function as secondary protection.

After various iterations, including a trial with me in a Sokol suit, they let me go and wash up. I remove the leftover plaster while they get the inevitable forms ready for my signature. My suit doesn't exist yet and my seat liner isn't ready either, but both are already expressed in numbers, pure and simple: suit no. 422, seat liner no. 650.

Just like so many of them, yet also completely different from all the others, these are the centres of nucleation around which my dream begins to crystallize.

—

During those weeks in Star City, I spent several more days in the hydrolaboratory training underwater, a very welcome opportunity, as always. This also gave me a chance to work closely with Sasha, commander of the Soyuz before mine. I already knew him fairly well, since over the summer we'd both taken the first part of the course on the ATV, ESA's cargo vehicle. During the first simulations of the rendezvous, when we monitored ATV's approach and docking at the Space Station, I'd had the chance to appreciate Sasha's professionalism. As operator number 2, he'd helped me manage the anomalies our instructors presented to us. This excellent impression of him was confirmed in the hydrolaboratory, where we were preparing together for the potential EVA that would see us installing the ERA as the back-up pair, in case it was postponed from Alex's expedition to ours. Delay was in the air, and a short time later it was actually announced, putting an end, unfortunately, to my training in the Orlan suit. As I'd feared, the launch of the MLM, and with it, the ERA, skipped right over Expedition 42/43, and was postponed to some unknown date in the future. Although not entirely unexpected, it was somewhat disappointing. The NASA suit was now my only hope of making an EVA in space.

It wasn't smooth going just then for the EMU either. For several months, it hadn't been cleared for use in orbit except in cases of absolute necessity. Meanwhile, the specialists in Houston were trying to determine the cause of a serious incident, when Luca found himself in a very dangerous situation. Of all the possible ways to die in space – and especially during a spacewalk – drowning has never featured high on the list, until the day when a large amount of water accumulated inside Luca's helmet, to the extent that eventually it stopped him from seeing or communicating via radio. Thanks to his professionalism and *sang-froid*, the experience of his EVA lead Chris and prompt support from Karen and his Russian colleagues in the ISS and Mission Control in Houston, he was able to get back to the safety of the Space Station, ready to underplay the situation with characteristic irony: 'Now I know how a goldfish feels!' I was in class in Star City while all this was going on. I read during a break that they'd called a Terminate EVA, and with the instructor's permission I followed the updates from Houston while my class continued. Actually, I didn't realize the gravity of the situation in real time: terminate doesn't mean abort; it does not indicate a serious emergency that requires an immediate re-entry. However, the incident was indeed quite risky, and they definitely needed to identify the root cause of the malfunction. If Luca's suit had experienced this unexplained failure, they couldn't rule out the possibility that another suit might manifest the same problem. And what if Luca had been working at the far end of the truss, a long way from the Airlock instead of close to it?

While a team in Houston was analysing the anomaly, methods had already been identified for mitigating the risk if the situation should repeat itself. It was clear that the water had got into Luca's helmet through the air vent level with the nape of his neck. From there, it had soaked through his Snoopy Cap and started to migrate towards his face, with the gelatinous behaviour of liquids in weightlessness, where surface tension takes over. So from then on, the back part of the helmet would be covered with an adhesive strip made of the same material as the nappy. Astronauts would in fact wear a type of large absorbent pad behind their heads. A snorkel would be provided in case so much water leaked in that the absorbent pad became saturated and water accumulated in the helmet. The snorkel would allow one to breathe air from the chest region, and it quickly became standard NBL kit, so that you could get used to it. As for the absorbent pad, I got acquainted with it during a dedicated session: while I was wearing the suit torso and helmet, the instructors poured greater and greater quantities of water on the pad to familiarize me with the increase in thickness I could anticipate depending on the magnitude of the leak.

In the meantime, investigation of the problem had located the proximate cause in an obstruction in one of the components, owing to contaminated coolant water. Further investigation over the coming months would trace the problem to an issue in the ground facility where the coolant water had been packaged. Various recommendations were put forward, and these were implemented before the EMU could be authorized for non-contingency spacewalks, more than a year after Luca's incident. The inquiry also highlighted the fact that the ground control team did not immediately recognize the nature of the problem or its seriousness, since that failure mode can occur only in weightlessness and was not well known. For me, it was a moment to reflect: after decades of honourable service and hundreds of spacewalks, the EMU could still offer up some surprises. It may have moved beyond its pioneering phase, but spaceflight is still far from being routine.

Training at the NBL continued as normal, and I had regular opportunities to practise with Terry in the EMU suit. But it wasn't only training that kept me busy in Houston during that sweltering August. When I managed to get away from JSC or the NBL on time, I fretted over much more trivial things, such as underwear and skin lotion. There were fifteen months to go before the launch, but my first baggage deadline for space was hanging over me: it was time to choose work clothes, toiletries and workout gear. For Terry and me, the process had begun months earlier with a trial session in Houston. We were shown various NASA products we could choose from. Trousers or shorts? Long sleeves or short? Socks or knee-highs? This barrage of questions was accompanied by an avalanche of information about how long we'd be using various items and feedback from past crews. There were also alternative Russian products, but we'd only see those some months later, which meant that comparisons had to be postponed. Bernadette was responsible for dressing the astronaut crews on the ISS, and in her methodical way she sent all the information ahead of time, complete with photos and suggestions. But who had time to read it all? So we found ourselves in front of a large table covered with clothes and gear, assailed by questions and confronted with a perplexing realization: it was more distressing to choose underwear than it ever was to spin in the centrifuge or deal with emergency simulations.

I asked lots of female veteran astronauts for their views, mostly when it came to choosing between different items of underwear, which for women are simply bought from shops. There was one question in particular which I'd never considered, yet it suddenly became relevant: did I need to wear a bra in orbit? If so, what sort? After listening to the advice of various expert colleagues, I opted for a bra camisole, which I felt would be enough for daily activities, and a sports bra for running. I also spent an unusual amount of time in shopping centres looking for face cleansers, moisturizing creams and foundations that didn't contain any of the prohibited ingredients, among them many commonly used alcohols. At the end of August, I could proclaim victory before going back to Europe: my list was ready for Bernadette. The first one, anyway. In the coming months, there would be many more lists and deadlines punctuating my way to space. I began to think that the training, which after all didn't require me to organize anything or make decisions, was the easy part of what lay ahead. What really began to intimidate me were the thousands of micro-preparations which neither Luca A. nor Alicia, his super-efficient counterpart in Houston, would put on my schedule.

As always, the training journey at JSC had been intense and taxing, made up of long days when lunch meant hurriedly buying a sandwich to eat in the next class. Back in Europe, I took a short break in Sicily and then joined Butch in Cologne for the course as 'specialist' on the Columbus lab. I once heard a colleague use the washing machine to explain the various levels of qualification on the Space Station systems: a user knows how to do the laundry; an operator understands all the functions and can interpret possible signs of malfunction; a specialist can take the machine apart and repair it. In some situations the analogy could be particularly fitting: one of my lessons as a specialist, in fact, was about looking for a water leak in Columbus.

Just like all non-Russian modules on the Space Station, the European laboratory is criss-crossed by a complex network of pipes. The coolant water running through that network draws heat from the equipment and then transfers it to the ammonia, the notoriously toxic substance in the external cooling loops. From there, the heat is transported to the Space Station's large radiators, to be dispersed into space. If there's a leak, the crew must work quickly to find and isolate the fault, before the risk of overheating the machinery requires switching it off. It's one of the procedures that definitely can't be performed by the controllers on the ground: the various racks have to be physically disconnected from the cooling loop until the leak is stopped, a sign that the faulty rack has been identified.

During the short time I was in Cologne, I was able to carve out a bit of time from my schedule for a few visits to the dentist. It's a myth that astronauts have perfect teeth; mine certainly aren't. However, I had never before experienced what happened to me next. A week after a minor procedure, I was still in pain. I stockpiled the pain relievers, and with the blessing of my flight surgeon Brigitte and my dentist's less than reassuring view that it was probably only a matter of another day or so, I set out for Star City. The next day I was scheduled to scuba-dive in the hydrolaboratory, and the day after that I'd have my first training session in the Orlan with Sasha. Toothache or not, I had to do it. Rescheduling something as complicated as a run in the hydrolaboratory is just as difficult as it is to delay the beginning of a training trip. The repercussions would have impacted my entire trip template, not to mention Sasha's and probably those of some of the other astronauts. Our lives were interwoven, and the closer we got to the launch, the farther the ripples from any change would spread.

The young Russian border police officer certainly caused some ripples the evening I arrived at the Domodedovo airport. He examined my passport long and intently and then, looking me straight in the eyes, he asked, 'How do you intend to enter Russia today?' With the valid visa in my passport, I thought. But something about the young officer's seriousness soon clued me in to the fact that he wasn't the one in the wrong. That evening, with thirty-six years to my name and more time spent in airports than at home in the past few years, I learned that granting a new visa cancels out the previous one. My new visa wasn't valid yet, and the old one was cancelled.

By chance Yuri Petrovich was travelling with me, and I called him for help. But his plea for them to allow me to enter Russia, by virtue of the fact that I was an astronaut, that I would soon be flying on the Soyuz and that anyway I was a good fellow, was unsuccessful. I'd felt momentarily hopeful, accustomed as I was to the idea that Yuri Petrovich could solve any problem. But I was taken into custody by an official and told that I would be put on a flight back to Frankfurt. I had my doubts about whether there was another flight that day, since it was already almost midnight. So there was a little time to find another solution. When it came down to it, this was Russia, one of those places where nothing can be done, but then, anything is possible. While the various officials discussed amongst themselves, I managed to get my boss, Frank, on the phone, and I told him what had happened – not without profound embarrassment. Frank alerted the ESA representative in Moscow, René, and he hurried to Domodedovo, contacting the airport consul on his way – a person with consular roles, or so I thought. The consul came to speak to me, and though he was very kind, he could only inform me – however contritely – that he could not allow me to leave the airport, alas, since without a visa I definitely couldn't enter Russia and at that hour he was unable to issue one. He did however wish me a good night in the transit area of the airport.

I resigned myself to spending the night half-awake on a seat with fixed armrests. I had a packet of biscuits and a stash of small bottles of water which Yuri Petrovich had somehow managed to get for me before he left. Thanks to the fax René sent to the Ministry of Foreign Affairs at four in the morning, and who-knows-how-many phone calls, I wasn't put on the first flight to Frankfurt the next day. At around eleven, I was able to leave the transit zone with a new visa in my passport. Nikolai and Yuri Petrovich were waiting to give me a hug. Of all the people involved in my mishap, I was undoubtedly the one who'd got the most sleep.

18.

Houston, 5 November 2013

I stop for a coffee at Starbucks on NASA Road 1: four lanes in each direction, becoming six at the junction, and a central reservation studded with palms.

Along the roadside, a series of low buildings with drab windows, and in front of those, small car parks. Small for Texas, that is. There's a drive-through, but as always the queue inside is shorter. I get back in the car with a take-away coffee containing as much caffeine as five espressos, I think to myself. I need it, since I'm jetlagged and scheduled for many hours of Prep and Post. Today we're going to simulate a day of EVA without the EVA itself.

From NASA Road 1, I turn into Saturn Lane and from there I take it right towards the main gate of the Johnson Space Center. I've been coming here for more than two years, but I've been in only a fraction of the many buildings spread across its 1,600 acres. They hold offices, laboratories, simulators, a clinic and many other things, both ordinary and extraordinary. In Building 31N, for example, the Lunar Sample Laboratory Facility houses 400 kilograms of lunar rocks, which were brought back to Earth by the Apollo astronauts. Mission Control Center, or MCC, is in Building 30. This is where flight controllers held their breath until they heard the historic sentence 'Houston, Tranquility Base here. The Eagle has landed.'

I park near the now familiar Building 9, a tall hangar, 200 metres long, with yellow tracks for the overhead crane running along its walls. Everyone calls it simply Building 9, but the correct name is Space Vehicle Mockup Facility. On the eastern side, full-scale mock-ups replicate the entire length of the ISS. You can go from the Russian Service Module, barely outlined, all the way to Node 2 at the other end, which is faithfully reproduced along with all the other NASA modules. I head for the Airlock and join the small group already gathered round a big table at the foot of the raised access platform that runs along the mock-ups.

It's been a few months since my last Prep and Post. Butch was in the suit then, and I took part as the IV, or Intra-Vehicular, the person who sees to all the operations related to the suit and Airlock before and after the EVA. Maybe one reason so many people want to do the spacewalk is that the alternative is to do the IV. The procedures are complicated, and you literally hold your colleagues' lives in your hand. Risking your own causes a lot less anxiety.

As usual, Regan has brought a big box of chocolate-chip cookies. To put it simply, I can say that I've learned everything I know about the EMU suit from him. It's been two years since my first classes, and by some strange coincidence he's still the instructor. At the time, he was in charge of my training as a reserve astronaut, and now he's been assigned to the EVA team for our crew, along with Alex and Faruq. Whenever I'm in Houston, we have regular lessons together here in Building 9, where, besides the Airlock, there's also a worn-out EMU suit which proves useful for maintenance demonstrations. Sometimes we meet upstairs in the Virtual Reality Lab to practise the flight with the SAFER rescue jet pack. And we often spend a few hours in a freezing cold, windowless room with a large schematics on the wall showing the suit's life support system, a complex collection of pipes, valves, reservoirs, pumps, fans, regulators, sensors and many other components that make the EMU a small, wearable spaceship. In that small room, Regan is at the controls of a simulation programme, which he can configure with all kinds of virtual suit malfunctions. For my part, I work on a DCM, the only part of the suit that's physically present: the Display and Control Module, an electronic box that's usually attached to the chest of the

EMU, fitted with switches and a small LCD display to monitor the telemetry and send commands.

Armed with a cuff checklist – forty small stiff pages, spiral-bound and worn on the forearm by astronauts in EVA so that information is within easy reach – I practise correctly diagnosing the malfunctions and following the proper recovery procedures, keeping in mind the mantra I learned during my pilot's training: 'Maintain aircraft control, analyse the situation, take appropriate action', in that precise order. It's little consolation if you manage to handle a failure, only to smash into a mountain or use up your fuel flying in the wrong direction. Yes, a spacewalk is more forgiving, since there is no risk of crashing or colliding with an obstacle, but there are always errors waiting to happen, ones that can aggravate a given situation. A simple tether snag, for example, is enough to make you lose precious time when you need to get back to the Airlock in a hurry. That's why we also train in the pool for suit malfunctions, even though the water suit doesn't have a working DCM. Every so often, as we work, an alarm goes off in the headphones, and Regan's voice announces the problem we have to resolve. In addition to the usual ones – like loss of pressure in the suit, excessive amounts of carbon dioxide or suspect voltage – there's a new one: water in the helmet.

I grab a cookie and head for the dressing room with Terry, where we'll put on the undergarments and the cooling unitard, or LCVG. The first three hours of today's Prep and Post have flown by, between introductory briefings and Airlock configurations. The moment has now come to put on our suits and practise the ISLE procedure. Pronounced like the English word, the acronym stands for In-Suit Light Exercise, the pre-breathe protocol normally used on the Space Station in the American Airlock. Though it's a lot longer and more complicated than the Orlan equivalent, because the lower operating pressure in the EMU carries with it a higher risk of decompression sickness, the ISLE established itself as the standard NASA method, replacing the protocols more often used in the past, which required you to pedal on a stationary bike or to spend the night in the Airlock at reduced pressure. To allow enough nitrogen to be released from the tissues, the ISLE has you breathe pure oxygen through a mask for at least an hour, while performing normal activities. This is followed by a partial depressurization of the Airlock down to 10.2 psi, or around two-thirds of normal atmospheric pressure, before you take off the mask, put on the EMU suit, fill it with pure oxygen and carry out fifty minutes of light exercise. The exact meaning of 'light' is what Terry and I are learning today. We're instructed to perform slow, measured movements with our arms and legs, as these serve to raise our metabolic rate a little and accelerate the release of nitrogen from the tissues as needed. But we mustn't get too tired, since during a real EVA we'd have to perform hours of exhausting work outside the ISS.

A real EVA. I want one so badly! I'd love to find myself in this situation in orbit, sealed in the suit, attached to one wall of the Airlock with Terry across from me, watching one another as we engage in a bumbling, slow-motion ballet. Today, the medical staff are monitoring our metabolic rates and they tell us to move with greater or lesser energy so that we stay within the expected limits. In orbit, we'd be trusting our muscle memory. In orbit . . .

I know very well that there's little likelihood of me making a spacewalk. At this moment, two EVAs are provisionally planned for Expedition 42, and if they are confirmed, Butch and Terry will carry out both of them. That's how it should be. None of us has experience of EVAs in space, but my colleagues have many more hours of training in the pool and have demonstrated better skills. In addition, the two EVAs are actually a single activity divided into two sorties, so it's entirely reasonable that they should perform both. And that's to say nothing of the fact that I need a size M suit, or rather torso. Right now, all the suits currently configured with the life support system are size L or XL, and no astronaut will need a smaller one until I get there. It's an added complication, because swapping the torso on a suit takes time, and on the ISS crew time is the resource in shortest supply.

I can't help hoping all the same. If I weren't hopeful, how could I continue preparing with the necessary dedication? Of all

the training, working in the suit is the only part I find really difficult, so I need to feel motivated in order to give it my best – out of
self-respect, if nothing else. And because, to put it simply, it's my
job. It's essential that I'm able to perform an EVA on the Space
Station, at the very least to provide the necessary redundancy,
in case Butch or Terry had a medical issue, for example. Many
of the critical components of the ISS are found on the outside,
so the ability to perform a contingency spacewalk in the EMU
should be maintained at all times.

And then, who knows? There's always a chance. And although
no one is ever entitled to an opportunity, opportunities are reserved for those who are prepared. Perhaps a third EVA will be
added to the plan for our expedition. In which case, Butch and
Terry will have acquired a lot of experience . . . and one of them
might be allowed to go out with a less experienced colleague.

—

Alongside the NBL runs and the Prep and Post classes there was an
occasional class in Building 9 with high-fidelity hardware, much closer to the equipment actually used in orbit than the pool mock-ups.
Sometimes we trained in the POGO system: suspended in mid-air
from a harness, we could experience the difficulty of maintaining a
stable position without the 'unfair' aid of water resistance. Working
with the PGT at high torque settings, for example, required a solid
grip on a handhold, so you wouldn't start turning around as soon
as you hit the hard stop. I enjoyed the all-round preparation, and
it seemed like my efforts were slowly bearing fruit. Another good
piece of news: the prototype for my tailor-made gloves was ready
for testing. After a long wait, I was about to have gloves that were
perfectly fitted to my hands.

Besides our EVA training, there was a wide variety of activities to
see to in Houston, ranging from official photos taken as individuals
and as a crew to simulations of water samples collection for periodic
microbiological and chemical analyses. In Building 9, sometimes we
discussed daily life on board, what the crew quarters are like, for example, and where to plug in the vacuum; other times we'd simulate

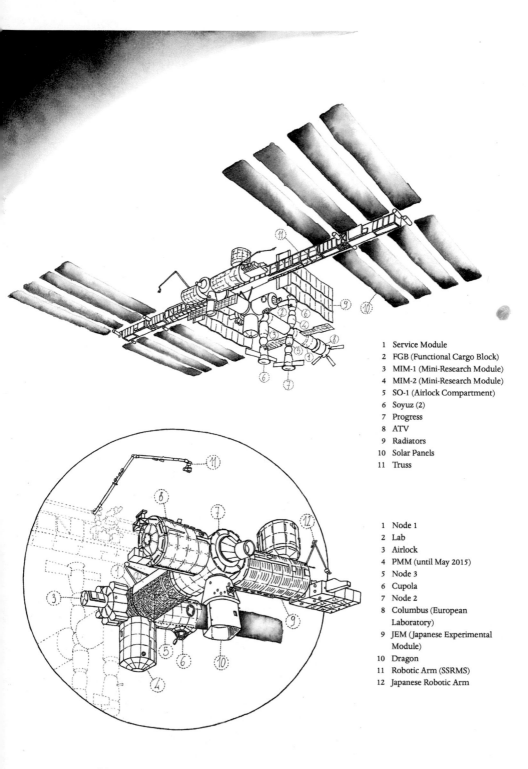

1   Service Module
2   FGB (Functional Cargo Block)
3   MIM-1 (Mini-Research Module)
4   MIM-2 (Mini-Research Module)
5   SO-1 (Airlock Compartment)
6   Soyuz (2)
7   Progress
8   ATV
9   Radiators
10 Solar Panels
11 Truss

1   Node 1
2   Lab
3   Airlock
4   PMM (until May 2015)
5   Node 3
6   Cupola
7   Node 2
8   Columbus (European
    Laboratory)
9   JEM (Japanese Experimental
    Module)
10 Dragon
11 Robotic Arm (SSRMS)
12 Japanese Robotic Arm

Top: Russian segment. Bottom: International Space Station (ISS)

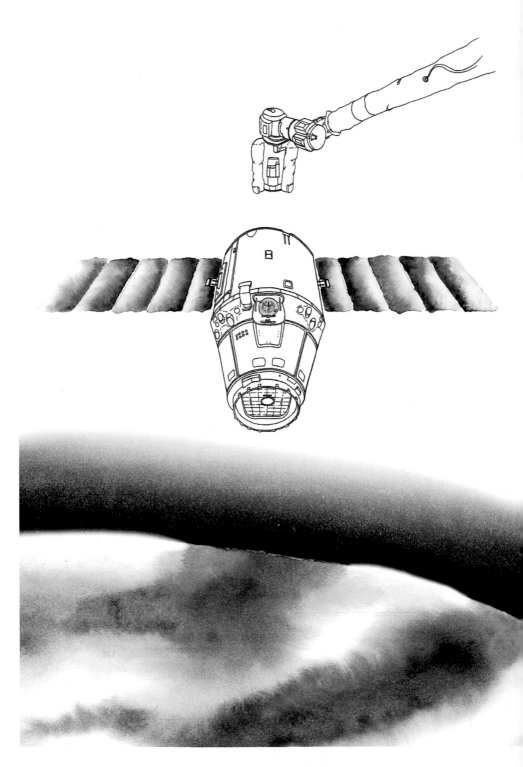

Dragon cargo vehicle

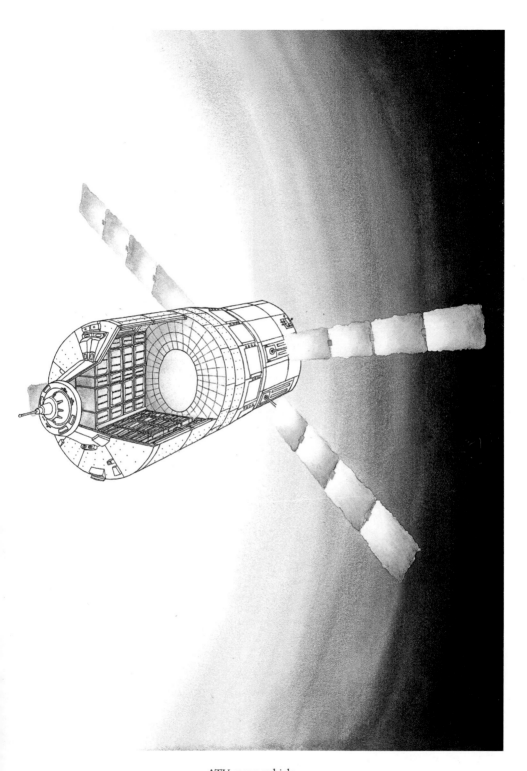

ATV cargo vehicle

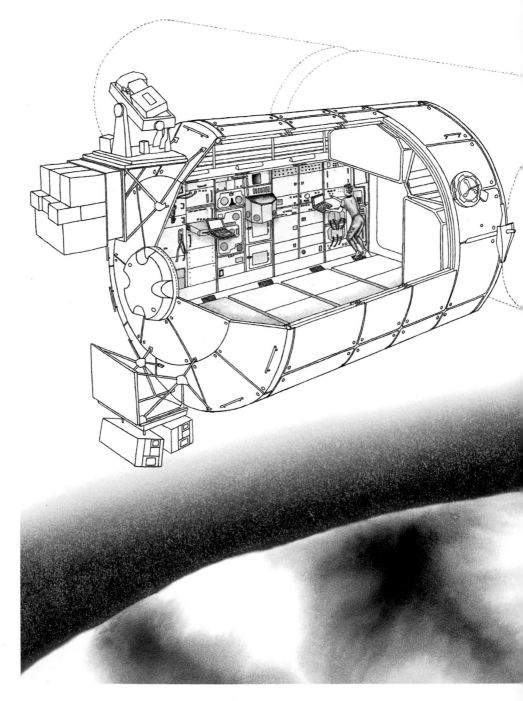

Columbus (European laboratory)

1 Docking and Attitude Control
  Thrusters (28)
2 Engine (hidden)
3 Infrared sensors

4 Periscope
5 Docking Probe (extended)

A Service Module
B Descent Module
C Orbital Module

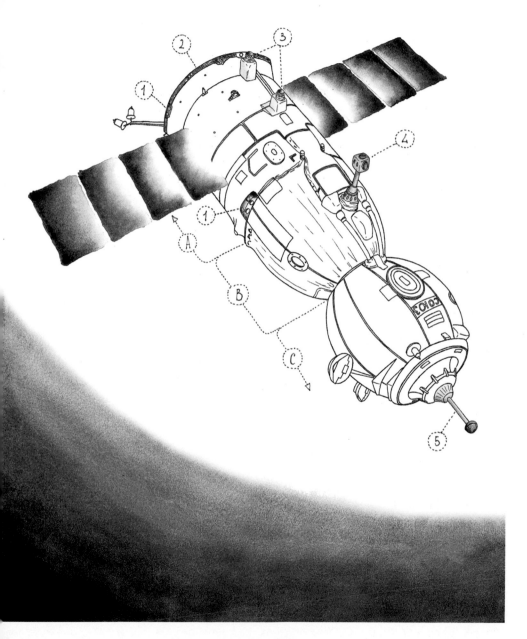

Soyuz spacecraft

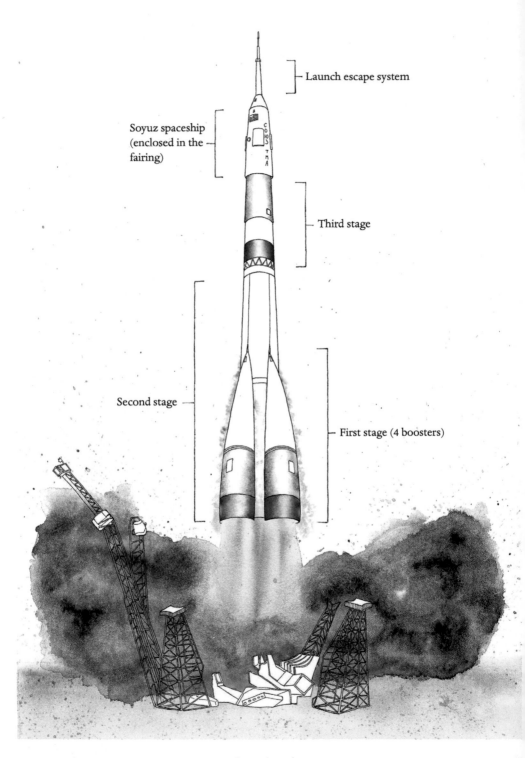

Launch escape system

Soyuz spaceship (enclosed in the fairing)

Third stage

Second stage

First stage (4 boosters)

Soyuz launcher

Orlan suit

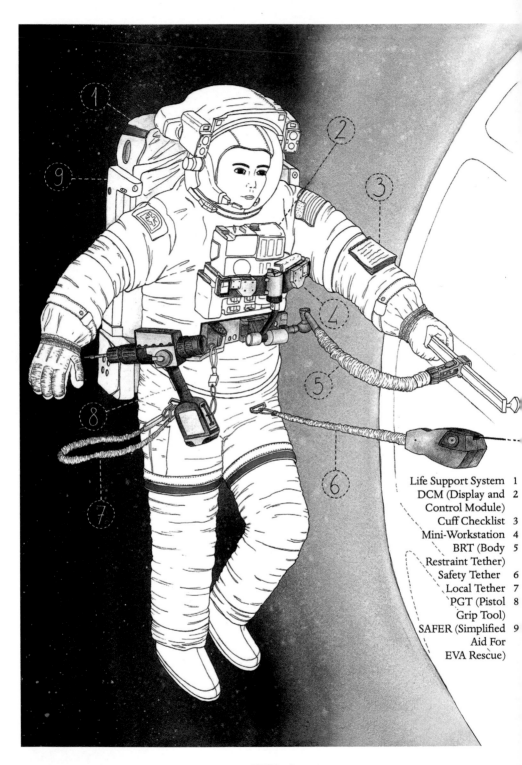

Life Support System 1
DCM (Display and 2
Control Module)
Cuff Checklist 3
Mini-Workstation 4
BRT (Body 5
Restraint Tether)
Safety Tether 6
Local Tether 7
PGT (Pistol 8
Grip Tool)
SAFER (Simplified 9
Aid For
EVA Rescue)

EMU suit

ISS robotic arm (SSRMS)

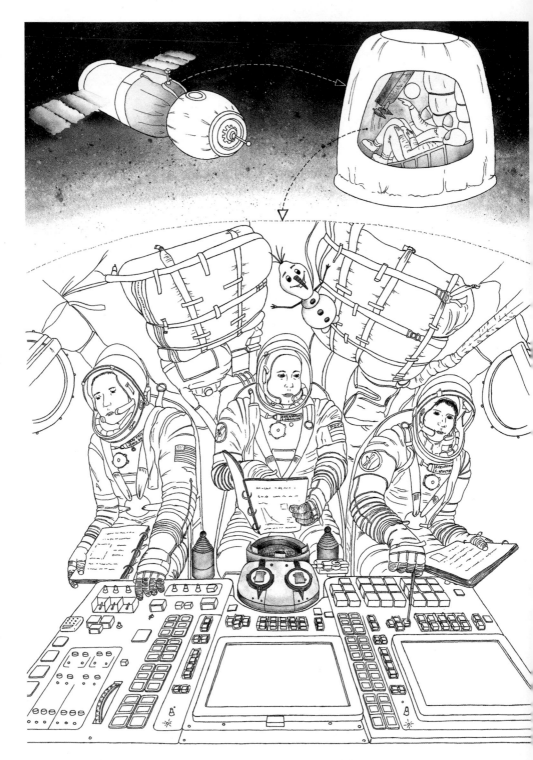

Inside the Soyuz descent module

Sokol suit

serious malfunctions such as the failure of a C&C MDM, one of three top-tier computers in the ISS command and control hierarchy.

On 7 November, Terry and I received an email from Baikonur with a peculiar subject line: 'Woooohoooo'. During the night, Soyuz TMA-11M had lifted off, and Reid had at that moment become the prime crew flight engineer. He wrote jokingly, 'We now humbly hand over to you the title of back-up crew.' Just six months later, in Star City, we would pluck from the table an envelope containing the scenario for our final exam as back-ups, and a few weeks after that, in Baikonur, we would leave Building 254 of the cosmodrome, one step behind Reid, Maksim and Alex as they headed for the bus that would take them to the launchpad, with family and friends shouting their final goodbyes. I was close enough to resonate with their emotions, but far enough away that I didn't feel the burden of responsibility, and I was possibly more moved at that point than when it was my turn to get on the bus.

The vast clockwork carrying me towards Baikonur was beginning to pick up speed, and the little army of people I regularly interacted with regarding Expedition 42/43 was growing ever larger. Brigitte was now officially my flight surgeon, and I appreciated her holistic approach to health issues. Deihiem, the ESA biomedical engineer assigned to the mission, would work closely with her. Biomedical engineers have a console position from which they can follow operations on the ISS in real time, attending to aspects that may not be strictly medical but nevertheless impact an astronaut's health, from air quality to sports activities to the daily workload. Alex, the ESA mission director, gave me regular updates on the mission content, often driven by changes in the launch plans for cargo vehicles scheduled to deliver new experiments or spare parts.

For some time now, whenever I was in Houston, I'd been receiving emails from Alonzo, who was responsible for making sure that Terry and I took all the many medical exams prescribed for the ISS crew. Alicia would put all our appointments on the schedule, but Alonzo always sent us a reminder, ensuring that we remembered any constraints, such as not drinking caffeine or eating, or anything regarding sleep or sports. Alonzo's work was starting to overlap with Matt's. Matt was the one responsible for the implementation of the human research programme, those experiments in which we astronauts act as guinea pigs, so to speak. In previous months, he had organized informed consent briefings for Terry and me, where the principal investigators had presented the experiment protocols and explained any possible risks to us. Having secured our availability and confirmed personal preferences, Matt had created our experiment package and now had to ensure that we took part in all the pre-flight sessions. Called the BDC, or Baseline Data Collection, these provide researchers with baseline values to compare with the data collected during the mission, so they can quantify any changes that take place as a result of being in space. Though it was unlikely that I would have to jump in for Reid in May, it was not impossible, so my first BDC took place already that November, a year before the planned start of my mission. As I would do many times before, during and after my flight, I wore a couple of temperature sensors for thirty-six hours, including two nights. One was stuck to my forehead and the other to my sternum, and the aim was to record fluctuations due to the circadian rhythm. The little suitcase containing the hardware also held a sombre black bandana you could use to cover the large sensor on your forehead and, at least partially, the cables connecting the sensors to the data logging unit, which you could fix on your arm or around your waist. As these BDCs began, and with increased medical exams, life presented me with something new to deal with. There would be weeks when I would have to wake up half an hour early to collect saliva samples according to a strict and laborious protocol, and days when I would have to juggle my other duties while laden with refrigerated bags for the collection of urine or faeces. When Terry invited me to go to a baseball game with him and Anton a few months later, I would take along a letter from NASA and a phone number for the head of security in case I was stopped at the entrance by stadium staff suspicious of the cables emerging from under my bandana and my T-shirt.

Anton's education and mine in American sports had already begun with a game featuring the Texans, Houston's American football team. It was a mandatory cultural experience, as much about the ads on huge oversized screens as it was about the game, not to mention the elaborate dances by attractive cheerleaders and gigantic glasses of Coca-Cola watered down by industrial quantities of ice. I hardly understood a thing about the game, despite Terry's dogged attempts to explain the rules. He relented when I assured him that, as a European, I also understood nothing about soccer. With this confession, my case could be filed away as hopeless without a second thought.

After Houston, early winter, cold and snowy, awaited me in Moscow, but first I would stop at EAC for a couple of weeks for a bit of office work and a few days of ATV training. With little more than a year to go before the launch, which had recently been moved forward to 24 November, I spent little time in Cologne now. I found myself checking the trip template to count the weeks when I'd be sleeping at home until launch: barely five. I was living my life out of a suitcase, and careening on tracks that led to the launchpad, where a rocket, bold and imposing, stood out in stark relief.

19.

Star City, 12 December 2013

When the alarm goes off, I automatically turn towards the Service Module Caution and Warning panel, a reflex I've acquired in all the previous emergency simulations. The light indicating fire is on, and the adjacent computer is already showing the page that allows you to determine which sensors have detected the smoke. At least two sensors have to be activated to set off the fire alarm. While I consider where the fire might be, Terry gets busy measuring the concentration of toxic products with a special instrument, which sits ready for use on the bulkhead. Since this is not a real fire, a value is suggested to him by Tanya, the instructor responsible for our training on the Russian segment of the Space Station. The carbon monoxide concentration is more than 200 parts per million: the air is toxic, and we need to put on protective masks.

The Russian masks consist of a large plastic hood in a military green colour, with a window over the face, and a little flask that hangs over your chest while you're wearing the hood. There are two openings in the flask: the upper one is connected by a flexible tube to the half-mask inside the hood, while the lower one has a black bag attached to it with a capacity of a few litres. In the flask is a chemical substance which, once activated, can enrich the exhaled air with oxygen by means of a reaction that consumes both the carbon dioxide and the water vapour. I remove one of the masks from the bag, unfold it, quickly check that it is in good condition, take a deep breath and put the hood on, still holding my breath. I pat it on top from my forehead to the nape of my neck to push out excess air, and then I tighten the strap around my neck, flip a lever to activate the flask and, finally, release my breath with some force in order to initiate the reaction. The black bag inflates and then deflates when I inhale again, and the air passes through the flask a second time before it returns to my lungs.

The air is warm and dry, with a strong aftertaste of something burned because of the exothermic chemical reaction. When I begin coughing violently, I start wondering if it was too warm or dry, or had too strong a burned aftertaste. It would be ironic to end up poisoned just as I was training to protect myself from that risk. It's difficult to know what's normal, since this is the first time I've activated the mask. In previous sessions we confined ourselves to putting on the hood and breathing the surrounding air through the tube that was disconnected from the flask. I look at Anton, the only one of us who's already had this experience, and he, too, is coughing, but he doesn't look worried. I'm probably not being poisoned. It's just one of those days when you have to put up with things. No one enjoys having difficulty breathing. Last night in Cottage 3, someone even described this simulation as the most unpleasant part of the training. But it's always better than having to acquaint yourself with the mask during a real emergency.

Visibility inside the Service module is slowly deteriorating as the air fills up with smoke simulant. We have to stay close to remain within sight of one another and we must carefully avoid crushing the black bag when it inflates with our expired breath: it contains our entire air supply, which is continually renewed and rendered breathable by the chemical reaction in the flask. There's enough reagent to last for about forty minutes and probably a bit more for a small woman like me – or much less if I were to breathe rapidly, as well I might in a real emergency.

We locate the virtual fire behind a panel and simulate an attempt to put it out with the extinguishers, foam ones from the Russian segment, which has a low-voltage electrical system. In the rest of the ISS, where the voltage is 120 V, carbon dioxide extinguishers are preferred – they stop a fire by displacing the oxygen it needs in order to burn. In actual fact it's fairly unlikely that a large fire would break out on board the ISS. Nearly everything is made of non-flammable material or kept in fireproof bags. And if the fire alarm is activated, it sets off an automatic response in the Space Station: the computers immediately shut down the ventilation and close the valves in the air ducts, isolating the segments in each module. In weightlessness there is no natural convection, the phenomenon which in our houses causes the warm air from our radiators to expand and become less dense than the cooler air around it, so that it rises to circulate in the room. Since the Space Station can't rely on this natural circulation, dozens of fans run continually to keep the air moving, so that air composition and temperature remain fairly homogenous across all modules. When there's a fire, however, all that takes second place to the need to extinguish it. By turning off the ventilation system, you prevent oxygen-rich air from being carried to the fire: once it consumes the oxygen in the immediate vicinity, it should go out.

It should, because the actual situation is not always predictable. In 1997, a malfunctioning oxygen generator cartridge caught fire on the Russian Space Station MIR, and the resulting flames spread so far that they blocked the escape route to one of the two Soyuzes and were hot enough that they could have melted the metal hull. While several crew members used the extinguishers, others prepared the one Soyuz still accessible for evacuation so that at least three of the six people on board could survive if it wasn't possible to get the fire under control. The fire continued to burn for several minutes, until the cartridge ran out, filling the MIR with thick smoke that forced the crew to wear protective masks for several hours. But everyone unanimously chose to remain on board.

Today we're training for just such an extreme case. Tanya announces that despite our simulated efforts with the extinguisher, we're not putting the fire out and – according to the scenario – it's spreading to the module on the Space Station where our Soyuz is docked, forcing us to abandon the ISS. In Star City, an evacuation means moving first of all to the building opposite, where the Soyuz simulators are kept. Our Sokol suits are waiting for us there, and we have to put them on without taking off the emergency masks except for the time strictly necessary. A brief interruption won't stop the chemical reaction; a long one makes the mask unusable.

So we're handed over to our Soyuz instructor, Dima, who's already sitting at the simulator console, ready to see that the fire follows us into our little lifeboat. From a long way off, I can see thick smoke simulant filling the descent module.

—

'Her name will be Futura,' proclaimed Lucio Dalla, the well-loved Italian singer-songwriter. I was humming his 1980s song one morning in Star City on my way to the manual-docking simulator, as I trudged through high snowfall from the night before. Though it was almost nine, the sun had not yet risen, and the thermometer on the balcony read -16° C. I'd heard the long-awaited news the night before: my mission would be called Futura. Officially, the mission would remain Expedition 42/43, but in keeping with established tradition for spaceflights with European astronauts, it would have a proper name on this side of the Atlantic.

I immediately liked the name Futura and was happy to share it with my small but enthusiastic group of followers on Twitter. Behind my happiness, though, was a muffled, almost imperceptible feeling. Now that it had been christened, the mission was making its society debut. Until that moment, it had been a fragile plant to protect and nurture, to keep out of the limelight as it grew up and became stronger. But the age of innocence was over. Many people would discuss it, many people would contribute to shape it, and I had less and less control over it. It was clear from the moment the name was chosen, and I had no part in that: Futura was no longer my little plant. It was now a tree and would be looked after by a host of gardeners. It was inevitable and somehow right that this should be the case.

By the end of the year, Futura had a logo, chosen in a public competition announced by ASI. It was a classic design, showing the ISS in orbit around the Earth, and it joined those of the

Expedition 42 and 43 logos, which had been proposed by their respective commanders, Butch and Terry. I found it very moving to see my name, along with those of my crewmates, embroidered on the patches that we all wore on our blue overall. There was only one missing – the one I'd help to create and with which I'd identify the most: the patch for our mission on Soyuz TMA-15M. A young cosmonaut suggested a unique design that established a clever parallel between the Soyuz and the attitude indicator, one of the main instruments in any aeroplane cockpit. Anton, Terry and I, all of us former military pilots, were immediately keen on that idea of celebrating the continuity between aviation and spaceflight.

Another friend, Riccardo, a graphic artist, had a clever idea for perfecting the artwork: he suggested that the Soyuz should throw a shadow in the form of an aeroplane over the Earth, and that the shape of that shadow would be made up of elements of the MiG-29, the F-16 and the AM-X, our respective planes.

While Riccardo was burning the midnight oil to finalize our logo, a short distance away, near Padua, someone else was labouring over another important part of our mission: food. The Space Station's pantry contains standard offerings, half of them supplied by Russia and the other half by NASA. On paper, the menus repeat every eight days and are designed to meet the crew's daily nutritional requirements. However, meal composition is not rigidly prescribed: for reasons both practical and psychological, we astronauts are free to choose our preferred packets from open containers, and we can boost the variety by swapping between the American and Russian provisions. In addition, before the mission, each crew member can fill nine large shoe-box-sized containers with their favourite foodstuffs to create a sort of personal larder, which we call 'bonus food'. In Houston and Star City, tasting sessions are organized in the years before the launch so that everyone can sample each dish on the ISS menu. As a result of these tasting sessions, a few packets of scrambled eggs and dehydrated asparagus would definitely be added to my bonus food containers. With NASA's approval, I would also add a few off-the-shelf products, such as dark chocolate, tahini and pumpkin-seed bars. But for the most part, my nine containers would be filled with dishes developed ad hoc by Stefano, a young and irrepressibly enthusiastic chef with a genuine dedication to healthy nutrition. I explained to Stefano that ESA had a small budget for producing a few new packets of space food to be included in the bonus food containers. Astronauts very often turn to famous chefs to create their countries' typical dishes, like Luca's lasagne or Alex's *Käsespätzle*. Stefano's mission, should he choose to accept it, was different: I wanted dishes built on modern principles of healthy nutrition. Stefano welcomed the challenge with humility and great determination and set to work at lightning speed to develop one-course meals that were tasty and nutritious and in line with NASA's stringent shelf-life requirements. I chose three of his proposed dishes: bean soup, quinoa salad with mackerel and cherry tomatoes, and chicken curry with peas and mushrooms. The packets would be sent to Houston in a few months and there packaged into bonus food containers for launch on ATV-5. My delicious quinoa salads would be on board the ISS a long time before I would.

The year was coming to an end. Exhausted by the continuous intercontinental travel, I had made no special plans for the holidays, but at the last moment, my partner Lionel and I decided to buy plane tickets, and we greeted 2014 at the opera in Budapest. We spent the first few hours of that long-awaited year quietly playing Risk in the apartment of Mercedes, who was a friend of my old friend Bernadett. The mood was jovial and familial. I happily gave in to the holiday warmth, knowing full well that in the year to come, moments such as those would be increasingly rare. A non-stop marathon had been set in motion, and it would continue right up to the launch. I wasn't that worried about it. By now I felt I was on a well-trodden path, and our training at this point was mostly review and consolidation; the logistics side of the preparations were also well under way. All I had to do was preserve my energy, work steadily and look after my health.

But the year 2014 had a nasty surprise in store for me.

20.

I have not been able to do these things because of any great
talent I possess; rather, it has all been the roll of the dice,
the same dice that cause the growth of cancer cells, or an
aircraft ejection seat to work or not.

   Michael Collins, *Carrying the Fire: An Astronaut's Journeys*

Moscow, 10 April 2014

The Sokol suit is never comfortable, that's for sure, but it's particularly uncomfortable when it's pressurized. That only occurs
in an emergency situation. During a nominal mission the suit
is only briefly inflated during the leak check, once before the
launch and again before separating from the ISS. For the rest of
the mission it stays deflated, unless there's a loss of pressure in the
descent vehicle. In that case, the suit, which is connected to the
on-board oxygen supply, maintains an internal pressure of 0.4
bars, which allows astronauts to continue breathing inside the
spacecraft even if they're surrounded by a vacuum.

   Unlike the suits used for spacewalks, such as the Orlan or
the EMU, the Sokol doesn't allow you to move much in pressurized conditions. When it's deflated, it provides some freedom of
movement and dexterity; when it's inflated, it has a single objective: to save the life of the crew, permitting only those movements required to complete the return to Earth. The outer shell,
which is usually soft, stiffens and expands to such an extent that in
the normal supine, crouched position the lower part of the back
doesn't make contact with the seat. Imagine yourself stretched
out on the floor, with your knees bent and your calves resting on
a table that's a little too high, so that your pelvis won't touch the

floor; imagine, too, that the edge of the table is very thin, and presses against the area behind your knees. That's usually the critical point: when it's inflated, the pressure of the suit on your legs can cause pain and hinder your circulation. If the suit is the right fit, however, there's enough space to lift your legs, one at a time, to relieve the pressure.

I know the situation well, since in February, I spent a good two hours stretched out in a Soyuz seat to test the comfort of my personal seat liner and my custom-made suit in pressurized conditions. Two hours is the maximum amount of time needed to perform an emergency re-entry following an air leak. Or a fire in space, for that matter: since there aren't any extinguishers on the Soyuz, the only way to put out a fire is to eliminate all the air by depressurizing the cabin. I would be the one doing that, since I would be the only one able to reach the critical commands panel, twenty-four large buttons under protective metal covers. When Anton confirmed, I would press command 20 followed by command 21: dual action for safety's sake. Once you initiate the sequence, there's no turning back. The air will escape into space.

Last February's fit check did not throw up any problems with the sizing of the suit, so today I am back at Zvezda headquarters for a functional test in a vacuum chamber. Aided by a friendly technician, I take my place in the seat and strap myself in, just as I would on the Soyuz: two lap belts, two shoulder belts and two knee straps. We start with a leak check. I put on my gloves and lower the helmet before closing the blue regulator valve on the chest. I have a simple pressure gauge on my arm, so I can observe increments in pressure and confirm that the suit inflates in the expected amount of time. The leak check is successful. The technician says goodbye and leaves the vacuum chamber, closing the heavy door behind him.

Before long, I can hear the pumps starting to work, and the virtual climb begins, towards the thinner layers of the atmosphere. We stop for a moment at 5 kilometres, or rather, at the pressure that corresponds to that altitude, around half the pressure at sea level. At this point, the suit's ventilation is arrested, and the oxygen feed is activated. The flow is much weaker, so much so that I immediately start to feel warm. As it happens, we'd have the same problem on the Soyuz. Two hours is considered the maximum time you can spend in the pressurized suit, precisely because beyond that limit the temperature would rise too high.

We resume our climb, and at 7 kilometres I see the needle on the pressure gauge start to move. The regulator has begun working: while the pressure around me continues to decrease, the valve keeps the suit at 0.4 bars. We stop at around 30 kilometres, where the surrounding pressure is about one-hundredth of what you experience at sea level: for today's practical purposes, this is vacuum. There's nothing more to do but wait, and try not to think about the discomfort behind my knees.

The conversation with the technician on the other end of the radio quickly peters out, and I'm left with my thoughts. I came back to Russia only the other day, after a slow orbit that took me to Japan, the United States and Europe.

Now that everything has turned out OK, I can stand back and allow myself to go over the events of the past few weeks, when I faced every astronaut's greatest fear before a launch.

—

It all began at the end of February while I was in Japan, training on the JEM laboratory module and for the Japanese experiment package to which I'd been assigned as a back-up. I was happy to be back in Tsukuba after almost two years, especially because this time it wasn't in the sultry summer heat. There's so much to love about Japan: toilets with more bells and whistles than an airplane; white-gloved taxi drivers; a rich and sophisticated cuisine like no other; elegant service; the Shinkansen bullet trains, which, as frequent as the Underground and timed to the second, cross the country at 300 kilometres per hour; sake and Kobe beef; onsens and the night climbs up Mount Fuji; Tokyo's great arteries flanked by skyscrapers – and then, just minutes away from the busy streets, peaceful village-like areas; noisy, crowded bars where the conversation resumes in seconds after an earthquake that rattles bottles and glasses – as if nothing had happened. More than anything else, I love the people's courtesy, something that is entirely Japanese and seems to correspond to a genuinely kind spirit.

After training one day, troubled by mild, intermittent pain in my abdomen, I asked to see a doctor. I'm not sure exactly what I expected, but definitely not what happened. Some hours later, when it was late evening, I found myself in hospital in Tsukuba having a CAT scan and a short time later I listened, petrified, as the JAXA doctors discussed the images with me. They showed a possible problem. This is the spectre that haunts every space mission: only nine months before the launch, and there was an issue that could compromise my medical certification.

Or maybe not. Had I misunderstood? I wasn't sure and I certainly couldn't take in all the ramifications. I was tired and cold, and everyone around me was speaking a language I didn't understand. Tomoko, a young JAXA doctor, translated into perfect English, but who knows how many nuances were lost on me? Saddened, I got into a taxi to return to my hotel. There were thousands of questions running through my head, and when I got back I immediately called my flight surgeon, Brigitte, who was still in Europe but already familiar with the situation. She tried to reassure me: maybe it was just a false alarm. In any case, the medical team would do everything in its power to allow me to fly.

In the space of a week, the little ache that had set everything off went away, and no one could really tell what had caused it. But by this stage, the lack of pain didn't matter. Even if I had no further symptoms, the possible issue revealed by the CAT scan would have to be investigated. Luckily, my next stop would be Houston, which probably has the highest concentration of world-class hospitals anywhere, and where NASA has a network of specialists who have taken care of other astronauts over the years, offering recommendations on their medical certification. After two weeks of ultrasounds and inconclusive lab tests, it was decided that, in order to secure the approval of the ISS medical board, I would have to undergo an invasive exam under general anaesthesia. It was the surgeon himself who shook me awake, post-anaesthesia, in order to give me the news I wanted to hear: he could now confirm with absolute certainty that there was nothing to stop me taking my spaceflight. Brigitte was there to share my relief, as was Dave, our crew's NASA flight surgeon, relentlessly optimistic and always quick to downplay the drama.

This ended the darkest month of my life as an astronaut. I was tired psychologically and I was tired physically, since I was on my final trip to Houston before the launch as a back-up crew member, and the training schedule was more demanding than ever. With her usual skill, Alicia had managed to carve out space for my medical visits – without ever asking an intrusive question. But the workload could only be rearranged, not reduced.

Now, just a few days after the good news, Terry and I had an important appointment: the EVA assessment. This is an NBL run observed by specialists from the entire community, who gather in the control room and for six hours note down every move we make, every word we speak in the water. At the end, a detailed report is drawn up evaluating each astronaut's readiness to perform an EVA in space. The gist of the evaluation, however, is relayed immediately during the usual post-run debriefing.

The underwater work went well – all the preparation and the workouts in the gym had paid off. By this point I was feeling more or less at ease in the suit and I had acquired good habit patterns that kept me out of trouble. I'd also learned to anticipate problems and to make clear and concise radio calls. Sure, there was still room for improvement – a lot – but the feedback was positive. Considering past difficulties, this success couldn't be taken for granted, and I owed it in great part to the coaching of Alex and Faruq, who always paid attention to detail and were never willing to lower their standards.

It was really gratifying to hear the positive comments – but my satisfaction was short-lived. At the end of the debriefing I was reminded that there was no ready-to-go EMU suit on board in my size and there wouldn't be one for the duration of the expedition. The message was clear: well done with your results up to now in EVA, but now please redirect your focus to the IV support activities inside the Station. This wasn't news. I was well aware of the situation, but hearing it so clearly and bluntly hit me hard, and took away some of my happiness at having done well in what was perhaps the biggest challenge of all my training.

After such a disconcerting few weeks, it was good to get back into Star City's reassuring routine, and I found it more comforting than ever. Elsewhere the unexpected could happen, but at the Profilaktorium I could always count on affectionate smiles from Yuri and Anna, and Dima was at the simulators, calm and unflappable, ready to refresh my skills in preparation for my imminent exams as a back-up crew member.

We had reached the final sprint on the long road to space.

21.

Climb up on the Moon? Of course we did. All you had to
do was row out to it in a boat and, when you were underneath, prop a ladder against her and scramble up.
  Italo Calvino, 'The Distance of the Moon'
  from *The Cosmicomics*

Star City, 11 April 2014

Relative to the Earth's surface, Soyuz and the Space Station
travel at around 28,000 kilometres per hour. But when we're
talking about speed during the rendezvous phase, we mean
relative speed, or the speed at which one is moving relative to
the other. At the moment of docking it is very low: 7 or 8 centimetres per second. When two metal shells containing precious
air touch each other in the vacuum of space, it's best that they
do so gently.

If everything goes to plan, the computer will make sure that
the speed and all the other parameters are correct. The Soyuz,
in fact, performs the approach and docking automatically thanks
to the Kurs system, a set of antennae that conduct a silent dia-
logue with their counterparts on the ISS, exchanging information about speed, distance and relative attitude. In the event of a
Kurs system or on-board computer failure, the astronauts must
take manual control. But there's a problem: the approach data
are no longer available. If this happens when you are not yet in
the immediate vicinity of the ISS but close enough to see it with
the naked eye, the flight engineer is expected in certain conditions
to move into the orbital module and use a laser rangefinder to
measure the distance and velocity. The drill for today, however,

concerns the final phase of the approach, when we are less than 400 metres from the Space Station. When you're that close to the ISS, all crew must be in their proper places in the descent module and the hatch to the orbital module must be closed.

But how do we determine the distance if the Kurs system isn't functioning? With the simplest aid possible: a grid attached to the viewer, superimposed over the image of the Space Station and linked to a table that correlates apparent dimensions and distance. If, for example, the diameter of the ISS Service Module is as large as a square on the grid, the table tells us that the distance is 200 metres; if the diameter of the docking hatch is two squares, the distance is only 70 metres, and so on. As for speed, the strategy is based on knowing the impulse given by the attitude thrusters, which are the ones fired during the final approach. An impulse of one second forward or aft means a 4-centimetre-per-second speed change. So if we find ourselves motionless relative to the Space Station and we want to approach at a speed of 40 centimetres per second, we have to provide a ten-second impulse forward.

It's only three weeks till the exam, so today I've asked my instructor, Sasha, if I can practise a sequence of four scenarios similar to those I'll have to deal with on the day. I take my seat in the middle and pull out the extendable track with the manual controls so I can reach the two knobs comfortably: the one on the right controls rotation on three axes, and the one on the left controls lateral translation, high-low and left-right. Next to the latter there's a lever. If you press it, you accelerate forward, towards the docking probe; you pull it to accelerate backwards. As we'll do on exam day, we start with the simplest profile, a change of docking port. The Russian segment of ISS has four possible 'parking spots', and although it's rare, you sometimes need to move your own Soyuz to make room for another vehicle. The change of docking port is always carried out in manual mode; there is no automatic sequence for this procedure.

The simulation begins as soon as the hooks open and, pushed by large springs, the Soyuz starts to move away from SO-1, one of the small modules equipped for Russian extravehicular activities.

Through the periscope, I watch as the docking port recedes and an increasingly larger portion of the nearby Service Module appears in my field of view. I wait until its diameter is reduced to four squares on the grid, a sign that it's about 50 metres away, before I give a three-second forward impulse to stop us receding. Now I start moving towards the end of the Russian segment to turn us around: I've actually left from a nadir position, or towards Earth, and I have to move to the MIM-2 module, which is at the zenith of the Space Station. When I'm in front of my destination docking port, I align myself with the target with a rotation of 180° around the axis of the Soyuz. The target is a diamond with one corner cropped and a projecting central cross, which shows attitude and alignment errors. I initiate the approach, gradually reducing the speed so there will always be enough space to brake. When I'm only a few metres away, my reference for the distance is the target itself. I want to stop at around 2 metres. It's a technique many astronauts use: stopping just before contact, and then giving a forward impulse for two or three seconds to acquire 8–12 centimetres per second: not too fast, not too slow, just the right closing speed to make contact safely. I've learned to take all the time I need and not feel tempted by the urge to complete the scenario. A few seconds of extra concentration can make all the difference. When I think I'm aligned, stable and still with regard to the docking port, I give a forward impulse for two seconds without further delay and continue to make the small corrections necessary to maintain proper alignment until contact. At this point the simulation stops. I relax my hands and wait for Sasha's feedback on the alignment, speed, time taken and fuel consumed. Everything was perfect. If this had been the actual exam, I would have received full marks.

A change of docking port is kind of a warm-up; more complicated scenarios are on the way. Sasha loads the next simulation, and I have a few seconds to note the speed and distance before the alarm goes off and the data disappear. I've lost the main navigation and control computer, and with it all the Kurs data and a series of automatic compensations that make manual flight easier.

This is the most difficult scenario. I turn off the alarm, activate the manual controls and quickly assess the position relative to the Space Station. Then I initiate the flyaround manoeuvre, a flight around the ISS to align with the docking port. We're about 300 metres from the Space Station. When I began my training I found it difficult to orient myself at this distance – the ISS components looked like a white blur through the periscope – but with practice I've learned how to interpret the image quickly.

In a few months I'll be seeing that whitish blur in orbit, and it won't be a simulation any longer, but a real destination. In the emptiness of space, it will be a friendly and familiar place, where fellow human beings will give us a warm welcome.

—

Tomorrow is Cosmonautics Day. For more than fifty years, the Soviet Union, now Russia, has celebrated the anniversary of Yuri Gagarin's historic flight on 12 April. It was the first daring orbit around the Earth, 108 minutes that ushered in the era of human spaceflight. Since 2001, the anniversary has also been celebrated in hundreds of other cities all over the world, with festivities called 'Yuri's Night', and in 2011 the United Nations declared 12 April International Day of Human Space Flight. Probably no one in Star City noticed. Cosmonautics Day has always been a special date here, and training comes to a halt so that everyone can celebrate.

On that 12 April no halt was necessary, since the anniversary fell on a Saturday. I would be celebrating Yuri in Cottage 3 that evening, but first I spent a lazy morning at the Profilaktorium with my Soyuz procedure checklists. After every launch, and often while I was away from Star City, updated pages were distributed; as always, Yuri Petrovich had left them on the desk in my room. So I made a big cup of tea and mentally prepared for a couple of hours of painstaking work, replacing the old pages with the updates, one by one. A few solitary flakes of snow drifted past my window, the last gasp of a winter that had seen temperatures of 30° C below zero. The first flowers around the lake announced that, at last, winter was giving way to spring.

I spent the rest of the day talking to my friend Michela via email.

We'd known each other since school, and now she lived in Vienna. I had turned to her for help since, as a PhD in English, she was well versed in literature, and I had embarked on a special endeavour: I'd been hoping to gather a selection of verse and prose extracts from various time periods and languages which would act as a sort of literary companion for my mission into space. I wanted to touch on themes of research and exploration, beauty and mystery, and our sense of awe in the face of them. The intent was to print this collection on minibooks, the kind that you can hide in your hand, barely 2 or 3 centimetres in size. I would be able to take a few hundred of these small, light booklets on the Soyuz as a part of my personal allocation of 1.5 kg, and when I returned to Earth they would make a lovely gift for family and friends, I hoped. I'd come up with the idea a few months earlier and I was still excited about it, but the Parisian publisher who'd offered to print these little books said that time was running out. That's how ideas and verses from Goethe to Emily Dickinson, quotes from the *Divine Comedy* and particularly lyrical phrases from Antoine de Saint-Exupéry started flying between Moscow and Vienna, as two friends exchanged ideas on Cosmonautics Day.

When it comes down to it, my first voyages were in books. I doubt whether I'd be an astronaut today if I hadn't climbed a ladder to the Moon many years ago, if I hadn't voyaged to the centre of the Earth, if I hadn't travelled all the way to China with Marco Polo or fought epic battles beside Sandokan. I lost myself in books, and found myself the same way. I've occasionally come across bad books, and I've learned to avoid them the same way you learn not to drink spoiled milk. I've also discovered wonderful books that have made me feel part of a shared human experience, made me feel small because I'm only an individual faced with the richness and intensity of many other lives, both real and imagined, but also large because I could identify with each of them through their stories. Books gave me words and imagination. I read them in secret, behind my desk at school and under the covers at bedtime, with a sense of daring and complicity. These were moments of solitude, but never of loneliness.

22.

Running on the beach is grand; learning geology is com-
mendable; jungle living is amusing; the centrifuge hurts;
the spacecraft ground tests are useful; but one is not to fly
until the simulator has told him he is ready.
     Michael Collins, *Carrying the Fire: An Astronaut's Journeys*

Korolev, 21 April 2014

The Russian Mission Control Centre, TsUP, is located in Korolev,
in the larger Moscow metropolitan area. I have only vague memories of my first visit here four years ago, but I immediately recognize the huge mosaic that looms over visitors in the entrance
hall. It shows the pantheon of cosmonautics: the visionary genius Konstantin Tsiolkovsky, who wrote the ideal rocket equation; the brilliant and clever chief-engineer Sergey Korolev, after
whom the city is named; and the young pilot Gagarin, who lent
his bravery and his smile to the Soviet endeavour to open the
road into space.

     We have a long day ahead of us. Upon our arrival, we are led
from the foyer to a large meeting room. Terry, Anton and I sit
down off to the sides, leaving the central seats for the prime crew.
In front of us sit approximately twenty specialists, each of whom
will discuss the mission with reference to their area of expertise. Initially, we speak about the Soyuz flight, and I learn lots of
interesting details, from fuel consumption to the Sun's angle of
incidence at the time of docking. These are preliminary calculations, and the final updates will be given to us in Baikonur a few
days before the launch. When the issues related to the Soyuz have
been covered, the discussion moves on to ISS activities and details

that regard only cosmonauts and their work on the Russian segment. I quickly lose interest, and my mind begins to wander.

This week will be dedicated mostly to revision in view of the coming exams, and the exams themselves will follow in rapid succession, beginning next Monday. I've already taken the one on the manual re-entry, which was brought forward to last week because there was some issue about the availability of the centrifuge. A great beginning – top marks – and a chance to get to grips with the formality of Russian exams. Before the next test, manual docking, we have a special appointment, one I'm eagerly awaiting: on Thursday, Terry, Anton and I will visit the headquarters of Energia, the company that manufactures the Soyuz. By rare and lucky coincidence, we'll have a chance to meet our spaceship, face to face.

Star City, 24 April 2014

My birthday is only two days away, and the gifts are already arriving.

This morning, when I woke up, I found an email from Brigitte reporting on the medical board meeting: I was fit for long-duration spaceflight, no questions asked. I have never been particularly worried about the yearly medical assessments, but after the contretemps of last month, it's a relief to know that my certification is now written down in black and white.

The most thrilling gift, though, is waiting for me here, in the big industrial warehouse where, after a slow gestation, the Soyuz vehicles and their cargo twins, the Progress, are born. Several of them are scattered around the building, some of them still in embryonic state, others at a more advanced stage of assembly. However, I have eyes for only one of them. Finally, the specialist who's leading the way points it out. It stands in a vertical position, surrounded by blue scaffolding: Soyuz TMA-15M. There it is, large as life, the spaceship that will carry me into space. It's nearly finished, and it's so beautiful, at least to my affectionate eyes.

Russian engineering is actually a distillation of functions pared down to their essentials, unembellished, yet with the genius of technical intuition. Formalism and individual flair coexist in a completely natural way, together with elements of craftsmanship that are rather touching: right next to our Soyuz, for example, homely women sit sewing the metallic fabric of the external thermal blanket. You would never come across the same scene in the ascetic clean rooms of Europe or the US.

There's a raft of white-shirted specialists waiting for us on the scaffolding. Some are holding papers or notebooks, probably procedures and documents to sign, yet many seem to have nothing in particular to do; maybe our visit is forcing them to interrupt their activity. On the third floor, Terry, Anton and I take off our shoes, put on a chef's hat and enter the orbital module on all fours. Things are familiar and strange all at once. It's not the simulator, worn by years of use. Everything's new here: the orbital module burns up as it re-enters the atmosphere, so nothing can be reused. Our homework for today is pretty simple: to check that there aren't any obvious discrepancies with what we expected. Aided by a list, we check each of the features we'll interact with during the flight, one by one, from the cargo straps to the carbon dioxide removal cartridges.

I move around the cramped space cautiously, taking care not to touch anything by mistake. It's partly sensible precaution, since soon I'll be entrusting my life to this assemblage of metal, pipes and electronics, and partly the behaviour of a guest who has the keys to someone else's house. The day when we can truly take possession of our Soyuz has not arrived yet. This is only a fleeting encounter, a chance to say hello before the spaceship undertakes the long train journey headed for Baikonur.

Star City, 28 April 2014

Today is the day Anton and I take our manual docking exam, which will test our ability to manually bring the Soyuz into contact with the Space Station using only visual cues. I haven't lost any sleep over it, but I'm fairly nervous. For some time, I've easily managed all the simulator scenarios Sasha's presented, but all the same, this is a tricky exam. Since I've been in Star City, I've known two veteran astronauts who've had to repeat it, despite being expert and well prepared, because the speed was too high at the moment of docking: the maximum allowed is 15 centimetres per second, little more than 0.5 kilometres per hour – any faster and you fail.

We've been in the simulator for more than three hours. Anton performed his four dockings, each of them perfect. We came out to stretch our legs for a few minutes, and then it was my turn to take the central seat. I completed the first two scenarios with full marks, and now I'm in the thick of the most difficult one: docking at the MIM-1 module without the main computer. MIM-1 is the actual module we'll be docking to in space. It's particularly tough because the target is rotated about 40° relative to the longitudinal axis of the ISS, so that you are continually correcting on two axes to compensate for the rotation of the Space Station. Yes – because the ISS is spinning as it rotates around the Earth: it brings the nose down, so to speak, to follow the curvature of the orbit and to maintain a constant attitude relative to the local vertical.

'TsUP in Moscow, do I have permission to proceed to docking?' I ask. I've completed the flyaround and I'm aligned with the port, about 70 metres away. Through my headphones comes Sasha's authorization to proceed.

In addition to managing the simulator and loading the scenarios, Sasha is also playing the role of Mission Control for our simulated radio communications. Behind her, on two rows of chairs, sits the examining committee, which includes representatives from Energia who've come here for the occasion.

I push forward the lever under my left index finger to start the final approach and correct the alignment with ever greater precision, as the target grows larger in my viewer. I stop, as always, at about 2 metres distance, and I take my time aligning myself perfectly. I have to give continual braking thrusts, because the thrusters I'm using to correct the attitude also have a forward impulse component, a troublesome coupling for which you must compensate manually if the main computer has failed. When I feel ready, I give a two-second forward impulse while I keep working to maintain the alignment, deflecting the controls in brief pulses. Shortly before contact, I have an unsettling feeling: I'm going too fast. At the last moment, I hit the lever backwards, lightly, to give a braking impulse. A very good decision, and just in time: Sasha announces a contact speed of 12 centimetres per second.

Relieved, I prepare myself mentally for the fourth and final scenario: night docking. I follow the instructions to approach up to 70 metres, switch on the floodlight and wait for orbital night. As soon as it's dark, I remove the solar filter from the viewer so I can see the illuminated target better. Flying is now more tiring. In order to keep the image within sight, I have to align my head perfectly with the periscope, and in the final approach, the light is blinding. When I stop 2 metres away, I'm blinking wildly, and I struggle not to allow my discomfort to rush me, since I know well that your sense of speed is distorted during a night docking. I need to be precise in halting the approach and then trust in those two seconds of impulse. And here I am, moving . . . a few seconds of holding my breath while I continue to correct the alignment, and finally . . . contact. At 8 centimetres per second, perfect!

I turn towards Anton, and we exchange a smile. Top marks for our crew. Maybe bad manners as far as the prime crew is concerned – there's an unwritten rule that the prime crew should be given a chance to do better than their back-up. But never mind! Maksim and Reid certainly won't hold it against us.

Star City, 4 May 2014

'Good morning, Terry. Anton says he's free at 2.00 p.m. Let's meet here at the Profilaktorium.'

'Perfect. My cottage is also an option. Big table, lots of space.'

Attitude: docking: LVLH + 180°

→ RS in flight direction w/
an angle of ~ 14° to provide
f. com via S-Band

undocking: LVLH

МБС на Союзе: необходима туда и обратно

___

1. 200 $ суток, аппарат СОТР
2. CO2 → 5,2
3. пров. СК в МИКе, 2,5 часа
    в ракету + осмотр + герм. СК ...
4. page 6 and further
5. check РЭАК / РОМК etc ...
6. РУЭ:
    - автом. на Вокурсе
    - ½ на 2-х сут.
    - вручную на фасетик.
7. Перекл. секций КДУ:
    в ЦВМ есть топливо, где
    идёт переключение.

My exam notes

'OK, let's go for your cottage. What number is it?'

'Five. Oh no! I guess I've never had anybody over!'

'Big day. You'd better do some cleaning!'

'Should I get dressed?'

'No. C'mon! It's not THAT formal!'

My email exchange with Terry has the usual easy-going tone, but the content is unusual: a Sunday study appointment with the entire crew. We don't often see Anton at the weekend because he's usually busy with his family, but I insisted that we get together today for a few hours. I've already spent an entire day going over the books with Reid, reviewing all the emergency procedures, but I'll be taking the Soyuz exam with Terry and Anton the day after tomorrow, and it's important that we're clear about the distribution of our tasks, above all in hectic situations such as a fire or rapid depressurization. Maybe it's not really necessary, but a little team chair flying won't do us any harm.

Up to now, the exams have all gone very well, and the final two were last week: the manual rendezvous, in which Anton flew four approaches from 3 kilometres to 100 metres away, and the one on the flight programme, an hour-long theoretical exam in which we answered questions about the sequence of operations and the content of communications with the Control Centre. The only ones left are two big simulations, each of them lasting a day: Soyuz and the Russian segment of the Space Station.

Star City, 6 May 2014

The morning started out calm, almost boring. Dima familiarized us with having up to twenty different failures in one simulator session; only five are expected on the exam. The first one came towards noon, more than three hours into our simulated flight. At 40 metres from the Space Station, a double malfunction with the Kurs system caused the computer to abort the automatic approach and to initiate a separation manoeuvre. The computer is programmed to do this for safety's sake, but our task is to interrupt that manoeuvre and to proceed with the approach in manual mode. While Anton was flying us in, I prepared myself for possible docking-system malfunctions, suspecting that we'd have to deal with at least one other failure before we arrived at the Space Station. The docking system has a suite of sensors and an ingenious working logic, but it's somewhat convoluted, and there are only seconds to recognize a possible problem and determine the action to be taken. Indeed: after the receiving mechanism on ISS already had a solid grip on our docking probe, a malfunction made the Soyuz believe it had touched the wrong place, so it ignited the thrusters for moving away, attempting to pull the 450-tonne Space Station along with it. We inhibited that automatic response too, and then continued to perform the procedure manually until we got the hooks closed. Now the lunch break is announced. 'War is war, but lunch is served on time.' Russians love to recite this saying of uncertain origin.

A van takes us to the cafeteria, perhaps once a sort of officers' club. There's a table laid for us, and at an adjacent one sit Alex, Reid and Maksim, who are today dealing with the exam on the Russian segment of the Space Station. They're curious to know which scenario we were given, since it means they won't have it tomorrow, but we don't have much to say yet. The more serious failures await us this afternoon, during our re-entry to Earth. After a fantastic borscht, followed by fish, potatoes and a couple of sweet pierogi – buns filled with cooked apple – it's time to return to the simulator and put on our Sokol suits again. I try to regain my concentration quickly, because the next half hour is critical, and we'll soon know if we've chosen an envelope with an especially challenging scenario.

We take up our seats in the simulator and prepare for departure from the Space Station, performing the leak checks and confirming the activation of the automatic undocking sequence. While the computer pressurizes the propellant tanks, I watch the telemetry attentively, ready to catch any indications of leakage: nothing, all is nominal. We are therefore ready to undock.

The smoke appears almost immediately, and right away we

realize that it's going to be a frenzied afternoon. Anton calls out, 'Close and lock your visors,' which our six hands are already doing in unison. It's my responsibility to lower the lever next to my left shoulder, the one that opens the suits' oxygen supply and halts the air circulation, preventing the smoke from contaminating our Sokols. A continuous flow of oxygen is necessary to allow us to breathe, to keep our visors from fogging up and to keep us from overheating. Yet it's also dangerous: only a small part of the oxygen that enters our suits is consumed by our breathing; the rest flows out into the cabin through the regulator valve, and this increases the flammability of the air in the descent module. We must depressurize and let all the air out into space when the oxygen concentration reaches 40 per cent, and not a moment later. It's a race against time.

As soon as possible, Anton launches Program 5, an automatic sequence that will set us up for a ballistic re-entry. I take Form 14 out of my kneeboard, looking for the right time to ignite the engine for this orbit. We have barely twenty minutes left. If we don't make it in time, we'll have to wait for the next orbit. An extra hour and a half in a depressurized capsule isn't a very tempting proposition.

Anton and I start to execute different procedures simultaneously, each trying to keep an eye on what the other is doing. Not checking each other's work increases the risk of error, but we don't have a choice in this scenario: time is too short. While Anton focuses on monitoring Program 5, I execute all the actions preparatory to depressurization. Terry, who's a whiz at mental arithmetic, keeps tabs on the partial pressure of oxygen and gives me an estimate of the time I have left to release the air from the cabin. When the moment comes, I turn a locknut and raise the cover on the two command buttons which, when activated in sequence, will open the relief valve in the descent module.

'Ready for depressurization?'

'Ready,' say Anton and Terry.

Command 20, followed by command 21. With the simulator's typical precision, the pressure gauge immediately begins to fall.

At this point, I rejoin Anton in the descent procedure. Program 5 is automatically taking care of the orientation of the Soyuz along the local vertical, but too slowly. While Anton and I discuss whether we should take over manually to speed up the process, a new alarm jolts us out of this uncertainty: the main computer has failed. We have to switch to the back-up computer, with its more limited capacity, and manually perform not only the orientation, but also the engine ignition and the separation sequence. At least we don't expect any further surprises: we've just managed the fifth and last failure, which put an angular rate sensor out of action.

Meanwhile, our suits are getting hotter and hotter. Luckily, a dozen procedures pages later, when we're virtually hanging from the parachute with only a few simple actions left to execute, the committee takes pity on us and declares the exam over. We have permission to leave the simulator and take a shower. I'm tired and hot, but really satisfied. I expect they'll point out a couple of minor mistakes, but I'm convinced that we did a good job overall. We are a well-honed team.

—

It wasn't quite time to celebrate – we still had one more exam the following day – but that evening I felt mildly euphoric. I was so happy: many people had come up to congratulate us. We'd made a fantastic impression. After a lifetime spent dreaming about going into space, I had got to the place where I was finally prepared. They were going to put me in charge of a spaceship, at least partially, in my capacity as flight engineer, in the knowledge that I was able to look after it. Wasn't this the most important goal? I was so pleased that actually going into space seemed like a minor detail now.

The next day, Terry, Anton and I showed up early in the Space Station mock-up hall for our exam on the Russian segment. 'Esteemed committee, the back-up crew for ISS Expedition 40/41 is ready for the exam,' Anton declared. I knew the routine by now. The six panel members, all of them space veterans, stood in a row in front of us. There were two NASA representatives: Suni, who'd come to Russia after her mission as commander of the ISS to run NASA operations in Star City, and Dan Burbank, who'd arrived from Houston especially for the exams. The last time Dan and Anton had been in this situation, they were on the same side of the table as crewmates. I suspect that Dan would have been happy to change places.

I selected one of the four white envelopes neatly lined up on the table, and all three of us signed it before climbing the stairs to the Service Module mock-up, where we waited for the simulation to begin. The envelope contained a series of malfunctions, and apart from one emergency scenario, they concerned Anton alone. For Terry and me it was a simple exam: all they expected from us was that we perform daily activities such as food preparation, using the water dispenser and the toilet, changing the container for urine and solid waste. I had to open and close some hatches, use various Russian computer screens, replace some filters and take a water sample for on-board analysis. As we expected, halfway through the afternoon there was an emergency scenario. A call on the radio alerted us that Mission Control could observe on the telemetry data a pressure drop inside the Space Station. We checked the unwieldy analogue pressure gauge – the instructors controlled it via radio according to the demands of the simulation – and of course we confirmed that the needle was moving towards lower pressures. We triggered the rapid depressurization alarm to initiate the automatic response from the on-board computer; we reset the communication system so we could receive Houston and Moscow on all channels; and we quickly moved into the Soyuz in order to ensure, above all, that our lifeboat wasn't the one leaking, and to calculate our reserve time, or the amount of time left before we would have to evacuate the Space Station if we couldn't isolate the leak. Retreating to the Soyuz for several minutes would have another advantage: it would allow the Russian airflow sensors to work undisturbed. With a bit of luck, they would automatically identify the affected module. They're not completely reliable, and for this reason we often simulate scenarios in which we have to find the leak manually, isolating each module one by one. But that day I'd chosen the lucky envelope: the airflow sensors successfully located the leak. Quick emergency resolution, exam over.

After a debriefing in a crowded room, it was time to celebrate the prime crew, who had meanwhile completed their own Soyuz exam. It was the first of many occasions over the next three weeks in which we would find ourselves gathered around a groaning table drinking toasts to Alex, Reid and Maksim with glasses of vodka and cognac: to the success of their mission, the love of their families, the professionalism of the instructors who'd prepared them, the strength of friendship in the international ISS community, and occasionally, to us, the back-up crew. Per tradition, we were jokingly urged to behave in such a way that no one would have considered us suitable for replacing the prime crew. This final toast was a relic of the past, when the back-up crew generally didn't yet have its own assignment. We did, though, and we were content to wait our turn, six months down the line. Far from envying my friends who were so close to launch, I expected the coming weeks to be some of the happiest of my life.

The festivities went on until late, and in fact the next morning we showed up rather sleepy to the meeting with the committee who would officially acknowledge our exam results. Formally we weren't ready to fly, not yet, but we were 'admitted to the next stage of training in Baikonur'. Only a few days before the launch, another committee would officially confirm us as prime crew and back-up. However, in actual fact only a serious and unforeseen event would keep Alex, Reid or Maksim from leaving for space. We would shadow them right up to the launchpad, and on the night of 28 May, we would watch them shoot into the sky in pursuit of the International Space Station.

After the meeting with the committee and a long press conference, it was time to take part in a series of ritual events. Like all the Space Station crews that had come before us, we, too, signed the visitors' book in Yuri Gagarin's office, faithfully reconstructed in the Star City museum, and placed red carnations on his tomb at the foot of the Kremlin wall. But I don't know if other crews had our good fortune. On that splendid spring day, the sky was clear, and Red Square was completely deserted. Access roads were blocked by barriers in preparation for the next day's celebrations: 9 May is Victory

Day (*Den' Pobedy*). We entered the square from the south, next to Saint Basil's Cathedral, its colours sparkling in the clear light as if it wanted to join in the festive atmosphere. I thought back to my first visit to Red Square on a grey, autumnal Sunday fourteen years before, when Saint Basil had been swallowed by a rainy sky and young women in red mini-skirts and stiletto heels braved the cold to sell cans of Coca-Cola to passers-by. At the time, I never imagined that one day I'd walk through a bright and empty Red Square accompanied by people with whom I was sharing the greatest adventure of my life.

23.

$$\Delta v = v_e \ln \frac{m_i}{m_f}$$

Δv = *change of the rocket's velocity*
v_e = *effective exhaust velocity*
m_i = *initial mass*
m_f = *final mass*

  K. Tsiolkovsky, ideal rocket equation, 1897

Star City, 15 May 2014

The meeting spot is outside, on a little road that cuts through the woods, in front of an unassuming old building. When Alex and I arrive with Yuri Petrovich, a small group has already gathered, and others arrive in dribs and drabs: men in jackets and ties, women in elegant dresses, a few little girls. Several have brought large bouquets. They might seem like guests at a wedding if it weren't for the fact that some of them are wearing the typical Star City khaki-coloured work uniforms. Looking carefully, I notice that many of the men have medals and honours pinned to their chests – the younger ones only one or two compared with all those worn by the older men, who are probably retired cosmonauts. The man with the most medals is Alexey Leonov, and he's unmissable even behind his dark glasses. He's a legend: in 1965 he made the first spacewalk in history, spending twelve minutes outside the Voskhod 2.

After fifteen minutes of greetings, hugs and small talk, the crowd moves inside for the breakfast ritual. Yuri Lonchakov, head of the cosmonaut training centre, quickly leads us in a series of mandatory toasts, ceding the floor in turn to representatives from various agencies and administrations, and other notable guests. Alongside the well wishes for the prime crew, there's a recurring toast addressed to us, the back-up crew: to be careful never to drop our guard. Of course not. God forbid! When it's Suni's turn to make a toast, she speaks from the heart. After reminiscing about all the time she has spent in Star City over the years, from back when she was a wide-eyed young astronaut, she recognizes how many of those present are now not only friends, but old friends. 'This is a fantastic place to grow old together!' she concludes emotionally. *Tri, chetyre* . . . hurrah, hurrah, hurrah!

After some twenty toasts, we are as usual invited to take our seats for a moment. If we didn't do so, there could be dismal consequences as we set out on our journey, according to the now familiar Russian superstition. Besides, this isn't an everyday journey: we're about to depart for the legendary Baikonur. Outside in the street, Terry, Anton and I walk behind Reid, Alex and Maksim, taking up our post for the next two weeks: faithfully following the prime crew. After the final group photo, with a monument of Lenin in the background, we're directed towards a waiting bus. There's a small crowd, some pushing, and Reid pauses for a moment to give his children one last hug, and then we're all inside, the door closes, and a short time later the bar is lowered behind us over one of the lateral access roads to Star City.

Two Tu-134 planes are waiting for us near the Chkalovsky airbase. Two, because we're not allowed to fly on the same aeroplane with the prime crew. Terry, Anton and I board the second one and we follow the prime crew fifteen minutes later. We make ourselves comfortable in the forward part of the cabin, a kind of first-class area with large armchairs arranged as in a living room. In line with tradition, and just like everyone else, I didn't eat anything at the farewell breakfast, so I hope we'll be having some food soon. I'm also curious: there are legends of a lavish meal being served during the three-hour flight to Baikonur. Many believe they load the breakfast buffet onto the planes, but probably it's only a way of answering the age-old question of what happens to all that untouched food after the crews depart. In any case, I wait patiently for the expected delicacies. At the back of the aircraft there are a couple dozen specialists from Star City in rows of seats configured as they are on a commercial airliner. These are the instructors, drivers, doctors, trainers and interpreters who will be spending our quarantine with us, the so-called operational group, or rather half of them, since the others are on the first aeroplane. They've already taken out vast quantities of drinks, fruit, cold cuts and preserved vegetables, but we politely refuse their invitations, since we are naturally waiting for a handsome banquet. From the pantry comes the sound of plates, and Anton reassures us that the flight assistant must be finishing her preparations. Anton should know: he's made this journey twice for his previous mission. But I think I can spot the Volga in the distance, a sign that we're about to leave Russia behind – and the promised feast has yet to arrive. A quick investigation reveals that our trip has begun with a minor catering problem. Could it be that someone failed to sit down at the departure breakfast this morning? Oh well, it doesn't matter. Our colleagues from the operational group are happy to welcome us, and we settle into 'economy class', eating well and laughing cheerfully. It's a close-knit group, and most of them have been to Baikonur together many times. Seeing them here, you'd think they were students on a school trip, but each of them performs an important function in the final preparations for launching Alex, Reid and Maksim.

With full stomachs and a few toasts behind us, I ask if I can take a look around the cockpit; or should I say, the nose of the aeroplane. This Tu-134 has a glass nose, as if someone had cut off the front of the fuselage and replaced it with glazing with the same shape. It contains an observation post that was formerly used for reconnaissance and visual navigation, and I reach it from the cockpit through a sort of tunnel, taking my place on the jump seat. It's like sitting on the edge of a cliff, but with no danger of falling. Below me is the Kazakhstan steppe, a monotonous flat brown landscape extending as far as the eye can see, and seemingly for ever: no matter how much we've left behind, the horizon keeps on spewing out the same amount of feature-less terrain. Human presence is rare and scattered, and yet it's in these lands that space exploration began. Huddled in the nose cone, I try to imagine what it will be like in the Cupola on the ISS. That, too, is a glass bubble of sorts, where you can admire the world. Who knows how different or similar it will be? We'll be weightless, farther away, going faster . . . Our journey today would take less than five minutes if we were going as fast as the Space Station.

I'm called back to the cockpit as we start our descent and I'm given permission to stay there until we land. At one point, the co-pilot directs my gaze to the Syr Darya, one of Central Asia's main rivers. Now that I know where to look, I make out Bai-konur beyond a bend, like a mirage on the arid, empty steppe. Once we've landed, we taxi to the parking spot and stop next to the only other aeroplane, the Tu-134 carrying the prime crew. Terry, Anton and I are the first to leave the plane, and we walk to-wards a small, orderly delegation on the apron, not far from the aircraft. After a somewhat formal initial greeting, during which we speak from a distance of a few metres away, the head of the delegation approaches us, smiling broadly, and shakes our hands warmly. They tell me his name is Sergey Romanov, and he's the head of the Energia operations here in Baikonur. He will oversee the final preparations for our Soyuz. My first impression is that he's an intelligent man with a kind heart, and I feel like we're in good hands.

The welcome ritual is brief: we take a group photo with the delegation and say a quick hello to local students, who've come with shiny garlands. We then head for the bus waiting on the apron. With us are Doctor Savin and Valery Korzun, the latter impeccably dressed in jacket and tie despite the sultry weather. Valery is a veteran cosmonaut who's flown on the Space Shuttle and Soyuz and visited both MIR and ISS, and he's now respon-sible for training in Star City. Here at Baikonur, he's the head of the operational group. I know him from all the exams; he poses as a severe and demanding evaluator, but I know that he'd throw himself into the flames to protect the crew. I've long had a hunch that his gruff exterior conceals a tender heart, a warm and compassionate attitude towards others. Doctor Savin is more inscrutable. Even before we met, I knew of his reputation through his nickname, 'Dr No'. His strict insistence on respecting the quarantine rules has been a source of frustration for more than one crew, but many have noted how he tries to avoid being anywhere he might find himself a casual witness to some small violation. Out of sight, out of mind.

We're relieved to find that the bus is air conditioned, since we're already sweating in our blue overalls. Reid, Alex and Maksim are not, of course, with us. We have to travel separately, and they're on the bus reserved for the prime crew. These are the same vehicles that will take us to the launchpad, and they have large seats with ventilation connections for feeding fresh air into our suits. The driver came with us from Moscow, and he's getting ready to ferry us to the quarantine, that mythical place not yet in space, but well removed from normal life on earth. All three of us are exhilarated.

After half an hour's ride, the Kazakh police car ahead of us stops in front of a big gate, which opens to let us in and then closes again immediately behind us, isolating us from the rest of the city. In truth, we won't be totally sequestered. We'll actually leave here tomorrow and for a second time a few days before the launch so we can go to the cosmodrome and carry out the Soyuz fit checks. Terry, Anton and I will also take part in the traditional visit to Baikonur, a privilege reserved for the back-up crew. A city of modest size, Baikonur has only about 10,000 inhabitants. On the road from the airport I saw mostly ugly apartment blocks in poor condition, some of them probably abandoned. But the theme of our trip will be space rather than architecture. I already know where we'll be stopping, having seen the sites countless times in photos with previous crews. We'll lay red carnations at the feet of the statues of Sergei Korolev and Yuri Gagarin, and we'll pose before the latter, imitating his gesture, immortalized in stone, of stretching his arms towards the cosmos. Then we'll visit the museum of Baikonur history and, on leaving, we'll take another classic photo inside a replica of a Kazakh yurt, wearing traditional clothes over our blue overalls. All I have to do in anticipation of those moments is to insert our faces in scenes I feel like I've known for ever. Of course, there are no sounds, smells or conversations in those images, no emotions and reflections. What will we feel? Curiosity, interest, pleasure, apprehension, fear, fatigue, impatience, excitement? I can't wait to become fully part of those scenes, and to experience every bit of life that links them up, all the things that you don't see in the official photos.

I must say I'm also really curious to hear the tales about Baikonur's history, so tightly interwoven with the early days of cosmonautics and the secrecy surrounding Soviet space missions at that time. When the city was built in the 1950s, there was nothing but the Tyuratam railway interchange. What is more, the name Baikonur belonged to a small mining city several hundred kilometres away, a trick to confuse foreign intelligence services seeking to locate the new Soviet cosmodrome. It's even alleged that fake structures were built in the real Baikonur to trick spy planes, but this may be only legend.

What isn't a legend, though, is what they say about quarantine, how pleasant the location and our daily life can be. We astronauts live with the Russian doctor in a small two-storey building situated in a lovely park full of leafy trees, which must be rather unusual in Baikonur's arid climate. I'm assigned spacious and comfortable accommodation on the ground floor with one bedroom, a large living area and a bathroom. The prime crew is on the second floor, and shortly after we arrive we all get together in Maksim's room to toast the beginning of the quarantine. Alex has already put an espresso machine in the hallway. He and Reid brought it all the way here, and will leave it as a gift to future crews. A few months ago, Reid even took orders for the coffee pods. Since Baikonur is three hours ahead of Moscow, our day is short, and it's soon time for supper. We eat in a small and elegant dining room. As Dr No explained to us, all meals are cooked specially for us according to stringent criteria regarding hygiene and quality. Obviously we're expected not to eat food from any other source, but I don't think this will be a problem for anyone, since the menu is rich and flavourful. Everyone knows that the biggest risk in Baikonur is putting on weight.

After supper we go for a walk in the park and we soon reach Cosmonauts Alley, a straight pathway paved with grey tiles and flanked on both sides by trees planted by all those who've flown into space from Baikonur. At the foot of the first tall and luxuriant tree is a sign that reads: 'This tree was planted by Yuri Alekseyevich Gagarin in 1961'. I don't know if it's true: I guess that if by any chance Yuri's tree had succumbed to the cold, aridity or disease, someone surely would have replanted it very quickly. I ask Anton to take a picture anyway in the soft light of the sun's last rays, and then we go to find his tree among the more recent ones. They're still young shrubs, planted alongside a lane that runs perpendicular to the main pathway, with the park on one side and, just beyond the row of small trees, a gentle slope towards the Syr Darya on the other. At the junction of the two roads is a small circular platform dominated by a model of the Soyuz rocket. From here, the view of the steppe is spectacular. We pause for a long time in silence, leaning on the balustrade and enjoying the cool evening, life's simplicity.

Baikonur Cosmodrome, 29 May 2014

The launch is scheduled for 1.57 local time. Fifteen minutes beforehand, we're taken up a short staircase, through a small opening, and out onto a narrow balcony that winds around the cupola of the small white building. This is the headquarters of the search and rescue teams, and it's from here that efforts to recover Reid, Alex and Maksim would be coordinated if there were an emergency re-entry during the ascent to orbit. Helicopters, aeroplanes and amphibious vehicles are deployed along the rocket's ground track – more than 5,000 kilometres – and the rescue vessel *Georgy*

Kozmin is on the alert in the Sea of Japan. At 2.06, we hope, they'll all be able to toast one more crew that had no need of their services. Everyone, that is, except those deployed nearby in Kazakhstan; they will remain poised to intervene should the Soyuz make an emergency re-entry after the first or second orbit. Tonight, I am conscious as never before of how the difficulties of running a human spaceflight programme go well beyond the construction of rockets and spaceships.

The rocket is over there, about 1.5 kilometres away, shiny and beautiful, plucked from the surrounding darkness by the powerful lights of the launchpad. The service towers, two shells that enfolded the rocket in a tight mesh of scaffolding and contain the lift that takes the astronauts to the top, are already retracted and lie horizontally. Only two umbilical connections remain, and those allow the liquid oxygen tanks to be topped up, slowly replacing the oxygen that evaporates and escapes as white puffs ruffled by the wind. The tanks' extremely cold temperature causes the water vapour from the atmosphere to be deposited in the form of ice, which now clothes the rocket in white.

The rocket was still green two days ago, when we saw it for the first time during the rollout, the transfer from the assembly building to Launchpad 1. An old train, for some reason painted in garish reds, blues, greens and yellows, took it away at a solemn pace. It looked like a train with human features, the kind you see in animated films, not modern ones but the older, hand-drawn films. A train with huge eyes for headlamps, panting from the effort and grumbling a bit, because it's done this trip for years and for nothing, not even a little thank-you. Everyone's here to admire the rocket, which isn't actually so impressive, lying there like that . . .

Actually, it doesn't stay down for long. Once it gets to the launchpad, it's erected in a vertical position in an astounding exercise of craftsmanship and visual estimation: 'Come forward – stop – a bit to the left – that's it.' We watched it, in disbelief, from inside the flame trench, a big circular ditch into which the bottom of the rocket is inserted to a depth of about 6–8 metres, and from which the burned gases flow through a special channel during the first moments of the launch. We were under the rocket, on a narrow service platform that runs right around the flame trench, while it was erected. And at the end, there we were, face to face with the twenty nozzles on the engines of the first and second stages, suspended before our eyes. Suspended: because the Soyuz wasn't standing on the ground, but rather held by the four arms of the launchpad, which wrap around it halfway up, like a ballerina in mid-air, her partner's hands holding her tight around her waist. When the rocket begins to lift off, these arms automatically open like petals without the need for any active mechanism, which could malfunction, but rather in a completely passive way, through the action of simple counterweights. There's nothing quite like the disarming ingenuity of Russian space technology.

The first umbilical is retracted, a sign that there's less than a minute to go before lift-off. Less than a minute, and we will no longer be the back-up crew. This has been a once-in-a-lifetime experience. We'll be back in Baikonur in six months, and everything will unfold in much the same way, but it won't be the same: it won't be spring, and we won't be so carefree. On the contrary, despite the ample amount of free time left by our work schedule, we will never feel like there's enough to finish all the things we still have to do.

We won't stay up late partying in the operational group's building, dancing until we're exhausted, and there won't be any late-night walks around the park after curfew, talking and joking under the starry skies of the steppe. The back-up crew can be forgiven a few small transgressions. A few days ago, late at night, Anton and I, along with a small group of young instructors, ran into the head of Roscosmos, the Russian space agency; he'd just come from Moscow and was taking a walk down Cosmonauts Alley. To our relief, the only thing he said when he saw us was 'Where's the third one?' as he looked around for Terry, who'd gone off to make a phone call. In Baikonur people like to see the crew together, even in the middle of the night, when they should be sleeping.

To mark the approaching end of our status as back-up crew, I popped into Anton and Terry's room this afternoon while the prime crew was resting. I pulled them along through a hole in the fence, all the way to Syr Darya. We're allowed to use that hole to go for a run, though probably we shouldn't go alone. We went to throw a bottle into the river, having stuffed it with a photo, one of thousands of crew portraits we had signed during endless sessions outlined on our daily schedule as 'symbolic activity'. We wrote a message on it and included an email address we created ad hoc. What a goofy thing to do! If everyone had the same idea, the river would be full of rubbish. My childish justification is that the occasion is really special and this part of the world is almost uninhabited. And luckily so, since every now and then a spaceship falls from the sky. I've no idea if some mysterious steppe dweller will one day send us an email, but the trip to the river gave us the chance to witness an amazing spectacle: a short distance downstream a herd of horses was drinking, a powerfully beautiful scene foreshadowing tonight's launch.

This is it, and I'm starting to feel some apprehension. I've already been to a launch, but that first time was very different. I didn't know a single member of the crew, not even Terry, who, by some twist of fate, is here beside me now. Today, at the top of the rocket sit three friends we hugged only two hours ago beside the launchpad. I don't think anyone can identify with them at this precise moment like we can, not even those closest to them, though they are certainly looking on with more apprehension. I can see them in my mind's eye, their visors lowered, eyes focused on the procedure checklist open to page 29, its famous diagram depicting the different phases of launch and the critical actions to follow in case of emergency. These are not discussed much during training, since the simulator doesn't replicate rocket failures, though they are explored in detail during the quarantine.

At the first flash of light, the sign that the engines have ignited, I press the call button on my phone, where I have already programmed Anton's number. We have decided to listen to his ring tone in the few seconds before blast-off; it has tormented us during our classes together, because of Anton's unquestioned habit of always keeping his phone on, but by now it has become the soundtrack for our crew. So, to the tune of 'The Final Countdown', we watch as the last umbilical is retracted, the engines reach full thrust, and a small sun lights up on the steppe accompanied by an explosion of noise and vibration. It's a gentle lift-off, and the rocket seems to pause, half swallowed by a white cloud of burned gases, as if entertaining second thoughts at the last moment. Then it stops stalling and rises, slowly but resolutely, perfectly stable in its trajectory, taking with it three fragile human bodies escaping the fate of being confined to the surface of this planet.

We rejoice, but only a little, because the ascent into orbit has barely started. After two minutes, we watch the first-stage boosters separate, four small lights that stay still and go out, while the small sun at the centre continues to burn and rise skyward, becoming ever smaller. Reid, Alex, and Maksim are soon just a small faint dot, which disappears from view. We quickly leave the roof and move to the operations room. All eyes are turned towards a screen that shows the internal video images, while the speaker broadcasts reassuring radio communications. After eight minutes and forty-eight seconds, we see them judder, the unmistakable sign of the jolt that accompanies the shutdown of the third and final stage of the rocket.

Reid, Alex and Maksim, despite the encumbrance of their suits and gloves, join hands in a sign of jubilation. They have become a new satellite of the Earth.

In the operations room the rescue team raise one of their first toasts to us. For the prime crew: *tri, četyre* . . . hurrah! hurrah! hurrah!

—

After that emotional highpoint in Baikonur, it wasn't easy to go back to the routine of training. To be frank, it was truly frustrating.

Inevitable, and expected, but frustrating. Four days after taking part in the launch, our eyes still shining with images, hugs and smiles, sparks of light in the night, dawn's flames reflected on Syr Darya . . . Only four days later I was in Cologne, at EAC for a week of refresher training on ATV. While my thoughts were in Kazakhstan, languishing in a fabulous dream from which I refused to wake up, and my heart jumped ahead, anticipating the cold and snow of Baikonur in November, a limp puppet sat in the simulator, mechanically executing operations that I knew inside out by now and for which I felt I needed no further training. It took me a while to come to terms with the frustration I felt about the next six months of ordinary life before my departure for space. Ordinary life. I don't know how many people would agree with that definition. Was I too used to a life of excitement, an extreme case of hedonistic adaptation? To be fair, Reid and Alex had warned us: they, too, had found the return from Baikonur disheartening.

Luckily, I was granted ten days' holiday. I'd known for a while that it would be my last chance to see family and friends who were normally far away, and for that reason I'd sent an invitation months before: Futura Party, 7 June, from 12.30 onwards. My parents organized a big party, blessed by a lovely early summer sun, which heightened the pleasure of seeing so many friends from childhood, adolescence and adult life. I was bound to most of these people not by virtue of seeing them regularly, but by an elective affinity that had survived infrequent contact and the fragmentation of my life, punctuated as it was with separations and new beginnings, a life in which I carried my house around with me like a snail. A slimy, silvery trail remained, and, looking back at it from this point in time, the whole of it traced and meandering behind me, it seemed to lead stubbornly and improbably towards a flight in space. That the snail would often take the wrong road, that providential rain arrived sometimes just in time, that other times a friendly hand had moved the snail a little further, taking it away from a busy road at the right moment – all that was impossible to see from a distance. Those faces and conversations brought back moments from the past, and they all seemed to contain an implicit promise of my inevitable future. I knew very well that this was an illusion, but just for that day I gave in to the irrational, and to the pleasure of feeling the past, present and future meld together in a dense, sparkling bead of quicksilver.

An unexpected flyover put the finishing touch on that beautiful day. Darkness was falling as we finished our meal, and someone received a text that the Space Station would shortly pass overhead. At the appointed time, all eyes looked skyward, searching for a bright spot moving west to east. I was the first one to spot it – or maybe they all indulged me in that particular privilege. We toasted to the Futura mission, the Space Station and the six humans on board. I felt alive: wholly and profoundly alive.

24.

Houston, 2 July 2014

My instructions are to wait inside Columbus. I can see Terry in the nearby JEM module, while my other four colleagues are scattered through the rest of the Space Station mock-up. It always takes a few minutes to get the simulation going, especially the first one of the day. The small army of instructors have to get organized: someone takes a seat at the simulator consoles at the foot of the mock-up near the Lab and Node 3; someone else puts on headphones to play the role of the control centres in Houston and Moscow; still others are spread throughout the different modules to observe our actions close up. We all wear a spy-mic so the instructors can listen in on our conversations. Today's dangers are all simulated, apart from the very real one of forgetting to turn off the spy-mic if I go to the bathroom. It must have happened before, because there's a sign on the door of the nearest toilet reminding visitors to turn it off. I've heard that there have been similar episodes on the Space Station, too: there are video cameras all over it, and occasionally one has been left on to capture and send back to Earth not only images, but sound as well.

The risk of embarrassment aside, it's rare for accidents to occur during on-board activity: in terms of occupational health and safety, it's difficult to do better than the Space Station. Not only is it impossible to fall off a ladder, but you're very unlikely to get an electric shock, since the high-voltage power distribution is made with extremely safe military-standard cables and connectors, and the voltage is in any case no higher than 120V. There are no pointed or sharp objects inside to bump into, nor are there any surfaces hot enough to cause burns. Potentially toxic substances used in experiments are rigorously isolated in three levels of containment, and the rare flammable objects must be kept in bags made of Nomex, a flame-resistant material. If any given activity could be dangerous, even if only slightly, the procedure will include appropriate warnings, recommending that the crew pay particular attention and exercise caution – to the extent that these sometimes make us laugh. Oh dear! The terrible danger of getting your finger pinched in a crack!

Humour aside, we're all conscious of how limited and precious our time is while we're in space, and we definitely don't need to be sidetracked by injuries or burns. Besides that, only minor medical assistance can be provided on board: any astronaut who was seriously injured would have to return to Earth, forcing the rest of the Soyuz crew to return as well. So it's natural to take every reasonable precaution.

Some situations are in a category of their own, since they can put the lives of all the astronauts in immediate danger. These are what ISS jargon calls 'emergencies'. Sometimes reality exceeds the imagination, and one day a crew may find itself confronting a danger that nobody had foreseen, or the probability of which was considered so remote that it could be neglected. It's possible, for example, that an asteroid could hit the ISS, but it's so unlikely that it's not worth worrying about.

There are, however, three emergencies captured in the procedures: fire, rapid depressurization and ammonia leak – contamination with the ammonia from the external cooling lines. I'll soon discover which of these is the first scenario our instructors have planned for us: I'm expecting the alarm to go off soon, and one of the three red warning lights to start blinking on each of the numerous Caution and Warning panels on the Space Station. But we're alerted to the emergency differently, this time: on the adjacent Node 2, I notice some smoke simulant. Terry is faster than I am to set off the manual fire alarm. The siren warns the others, and the computer responds by shutting off the air circulation.

If we could see where the fire had started, Terry and I would use the extinguisher right away. But the smoke is coming from behind the front panel of one of the racks, and it's not immediately clear where it started. So we need to rejoin the rest of the crew, and in order to pass through Node 2, which is now filled with smoke, we each grab a protective mask. In the Station's non-Russian modules the oxygen masks are practical and light and can be donned in a few seconds. They definitely don't last as long as the Russian masks, with their rebreather system, but they provide around eight minutes of continuous oxygen supply, with a slight delivery overpressure that prevents toxic combustion products from getting in. For today's purposes, it's enough to act it out: we put on the mask, open the valve on the empty oxygen bottle, fix it around our waists and move towards the stern, taking care to lower the hatch between Node 2 and the Lab in order to limit the spread of smoke.

In emergency situations, the crew gathers in the 'safe haven', a place on the Space Station that is not in immediate danger and offers easy access to the docking locations of the Soyuz spacecraft, which are our lifeboats. Conditions permitting, the safe haven is usually the Russian Service Module: the Soyuzes are close, the command and control computers are always on, there are paper copies of the emergency procedures, and various radio panels that can actually be reconfigured from there for simultaneous communication with Moscow and Houston. Also, the portable instruments that allow us to measure the concentration of toxic combustion products, starting with carbon monoxide, are on the walls, ready for use. Yelena and Anton have already taken the measurement: the atmosphere here isn't contaminated, so Terry and I can take off our masks.

The Service Module is really crowded. I'm not used to that. Already this morning, when we gathered at eight o'clock as usual around the big table at the foot of the mock-up, there wasn't enough room for all of us to sit. As with all simulations of this sort, there are at least ten instructors, ready to answer questions on their particular systems. But today there are also unusual visitors such as our flight surgeons and a couple of flight directors.

And then there's us, the entire crew of Expedition 42. The six-person emergency simulation is an important event, and the trip template is built around this hard requirement in order to guarantee the simultaneous presence of both Soyuz crews in Houston. Today Terry, Anton and I are training for the first time along with Butch, Yelena and Sasha. Next week, we'll do it all again with our colleagues from Expedition 43.

All of us have taken part in lots of similar training sessions in groups of two or three, and by now we're quite proficient in executing emergency procedures. But with a group of six the dynamic can be quite different. Being able to count on more people can definitely be an advantage, but only if we don't get in each other's way: it's essential for us to communicate effectively and be clear about the task distribution. Responsibility for coordinating the crew's emergency response rests on the shoulders of the Space Station commanders and is their one distinctive role onboard. During everyday activities, decisions are taken on Earth by the flight director, the person ultimately responsible for operations success, but in case of imminent danger, the commander is charged with both the duty and the authority to take the decisions necessary to guarantee the safety of the crew. So before the simulation began, Butch took time to explain what he expected of each of us in the various emergency scenarios. Accordingly, Sasha and Yelena are now to retrieve the respirators and put the fire filters on them, and then Terry and Butch each take one and go forward – towards the bow – so that they'll be ready to locate and extinguish the fire. It's my job to tell them where to look for it.

I take my place in front of a computer, open up the procedure for locating a fire, and start meticulously analysing every clue that could be useful: smoke detectors in alarm, circuit breaker trips, failed equipment. I try to be quick but not hasty. For most of the procedures, it's enough to know how to read and execute, but in this one we also have to know how to interpret and prioritize. When I think I've found the rack that's concealing the fire, I send Butch and Terry the identifying code for the closest fireport so they can insert the appropriate probe and measure the concentration of combustion gases behind the front panel. When the location of the fire is confirmed, it's my turn to get to work once more. Armed with the correct procedure and a bit of patience, I send shutdown commands for all the electric equipment in that sector. If the fire should continue regardless, there's nothing left to do but to discharge the CO_2 extinguisher inside the rack, through the fireport. Carbon dioxide, an inert and electrically insulating gas, will put out the fire by cooling and smothering it.

—

With everyone from Expedition 42 in Houston at the same time, there was an opportunity to complete a project that greatly appealed to the geek in me: the poster for Expedition 42. An exuberant ISS tradition, the expedition poster sets a photo of the crew in a lively graphic composition often inspired by a film poster. Having received carte blanche from Butch, I proposed to my crewmates *The Hitchhiker's Guide to the Galaxy*, with its famous reference to the number 42 as the answer to the ultimate question of Life, the Universe and Everything. The film poster lent itself perfectly to representing our six-person crew: there were in fact five people, and one of them conveniently had two heads. I was by no means the only one enthusiastic about this idea: Sean, the graphic designer assigned to our project, had recently written to me in the middle of the night while watching the film – 'in order to get ready' – and Glenn, one of our instructors, along with his wife Melissa, had been working for months on costumes and props. The evening before our photo session, I went for a costume test in their house, which was overflowing with children, animals and set designs fashioned over the years for parties and plays. Glenn's latest creation, which I could envision soon hanging on a wall, was a copy of the Point of View Gun – an imaginary weapon which, in the film, forces the person being shot to see things from the shooter's point of view. Melissa had prepared the costumes for each character, sewing some of them herself and then scouring the local second-hand shops. The next day she arrived at the photo lab before we did, ready to direct operations. It was priceless seeing Terry with his flamboyant mane of blond hair and Sasha posing in a blue tunic and a pair of rose-tinted John Lennon specs. Then I saw Butch, and I couldn't believe my eyes. Dressed as Arthur Dent and holding a cup of tea, he stood in front of the mirror trying out his best dazed-and-confused expression, with the same concentration I imagined he applied to landing on aircraft carriers. Douglas Adams would have approved.

Approval, however, was not forthcoming from NASA's legal office. To my great dismay, I learned that we wouldn't be able to make our poster public unless we obtained permission from the owners of the film rights. Easier said than done. After weeks of fruitless attempts to communicate with them, it truly seemed that our project was to flounder over an issue of intellectual property. *Dura lex, sed lex*. Until one day, Marsha Ivins, a veteran astronaut of five Space Shuttle missions, happened to hear one of our frustrated conversations on the topic and, incredulous, asked us why we hadn't spoken to her before. Didn't we know that she'd been friends with someone from Douglas Adams's family for years? Ah! How could we have missed that! I decided that in future, when in doubt, I would always call Marsha with any insoluble problem.

Though she'd left NASA some time ago, we often saw Marsha since she now worked as a consultant on an IMAX film that would be shot on board the ISS during our expedition and the one immediately after ours. Toni Myers was the driving force behind the project. She'd written and edited IMAX films shot in space since the early, atmospheric *Mission to MIR* and *Destiny in Space*. I'd seen them as a kid at Space Camp and found them fascinating: the screen size and the resolution, both larger than in conventional films, offered an involving and spellbinding experience. Thanks mostly to Butch and Terry's tireless camera work, *A Beautiful Planet* would come out in April 2016, showcasing the beauty of our planet in IMAX cinemas throughout the world and discussing the enormous challenge of keeping it habitable for future generations. It also talks a little bit about us and our adventure on the Space Station.

For Yelena, Sasha and Butch, the adventure would begin in less than two months, and their last training period in Houston was punctuated with traditional events such as the souvenir photos with our instructors and the cake-cutting ceremony featuring two huge cakes decorated with the logos for Expeditions 42 and 43. Butch and Terry followed one another at the mic to say a few words of thanks to those present, all those who in one way or another had assisted in the preparation of our mission. We lingered for almost an hour talking, thanking people and saying our farewells.

It was soon time to say goodbye to Butch. He was leaving Houston for two weeks' training in Europe and Japan, and until his launch we would be missing each other in our travels around the world. Hearing him say, 'See you in space!' had a huge effect on me, because it's not every day that you arrange to meet someone off the planet, but also because along with the promise of an imminent voyage, that farewell signalled that an exceptional experience was coming to an end. With these first goodbyes, the strange life I'd led over the past few years – stretching across three continents – was beginning to unravel. I would miss it: never before had I felt so completely at ease, stimulated and supported all at once, full of trust and admiration for the community to which I belonged.

It had been a privilege, for example, to get to know the women of the NASA astronaut corps and become friends with many of them, sharing something more than the camaraderie I felt with my male pilot and astronaut colleagues. I can't really say that working in overwhelmingly masculine environments has ever posed a problem for me. On the contrary, my character and inclinations are probably particularly suited to such environments in ways that I sense but would find difficult to explain. And aside from the EMU suit, which was obviously not designed with the anthropometric measurements of an average-sized female in mind, I've never encountered any obstacles that were clearly linked to my being a woman. In hindsight, even the impossibility of getting into the Air Force Academy after high school seems like good luck, since it allowed me to take a degree in engineering and to forge rich international experiences before beginning my military career.

Yes, I believe I have observed time and again that women are less easily forgiven any form of mediocrity – although by definition most of us human beings are mediocre at most things. But I have never encountered any intentional gender hostility – at most, the occasional paternalistic attitude on the part of men from older generations. I should make it clear, though, that I don't consider gender to be irrelevant: it's one of the key features of our identity, in most cases it's immediately obvious, and in many situations, above all in our private lives, it shapes our relationships and expectations. In my view, it's of course desirable to pursue the ideal of treating men and women equally in professional contexts. But even if we assume good will and acute self-awareness, it is just that: an ideal. So it's safe to assume that I have probably experienced occasional discrimination that was subtle enough to have been ambiguous. And it's equally likely that on occasion I have been subtly favoured because of my gender. Given that there's no way to measure it objectively, and considering it pointless to brood over vague suspicions, I've always found it practical to assume that, on balance, I've come out even.

What I didn't know, since I'd never previously experienced it, was how pleasant and reassuring it was to have more experienced female colleagues to turn to for advice or encouragement. I would miss being around them. I was thinking about just that one day, feeling grateful and already somewhat nostalgic, while on my way to a Ladies' Astronaut Night, a party for women astronauts that had been organized as a farewell for me and Yelena in view of our forthcoming departure for space. From Anna Fisher, of the legendary class of 1978, all the way to the most recent selectees of 2013, four out of a group of eight, over thirty years of female astronauts were represented. One of these was Peggy, who had invited us to her house. A few years later she would go back to space for her third long-duration flight, and her second as commander of the Space Station. When the Russians decided to launch the next Soyuz with a crew of only two people, opening up a seat for her on a later return flight, Peggy offered to stay longer on board, accumulating a good 290 consecutive days on the ISS. Today, she has spent more total time in space than any other non-Russian astronaut and has performed more than ten spacewalks. It was whispered in the corridors of JSC that Superman's outfit was just Peggy Whitson's pyjamas.

Towards the end of July, Terry and I also left Houston for Moscow, where we would be refreshing our skills on the Soyuz. A few days later, in the middle of the Russian night, an Ariane 5 rocket launched from Kourou in French Guiana, sending ATV-5 Georges Lemaître into orbit. It was carrying seven tonnes of provisions, ranging from propellant and oxygen to the so-called dry cargo – equipment, spare parts and consumables for the crew such as clothes and food. Nearly all my containers of bonus food were now in orbit. I wasn't surprised when I woke up to find an email from Stefano, my chef and by now my friend. He must have been up late watching the departure of my precious bonus food packets, which he'd worked on so enthusiastically for over a year. His message was bubbling over with joy, but above all, gratitude. In the years to come, I would discover that that was just Stefano. He experiences everything as a gift and is always finding reasons to be grateful. If that's a disease, it's one I'd be happy to catch.

ATV-5 docked only a couple of weeks later. In the meantime, with the aim of testing a suite of laser sensors and infrared video cameras, it carried out a manoeuvre that saw it fly barely 5 kilometres beneath the ISS, overtaking it. One lovely summer's evening in Star City, we watched both of them cross the sky while we sat outside Cottage 3, chatting around a fire: the Space Station and, just ahead of it, the smaller, fainter light of ATV. I was reminded of an evening four years earlier, when, by chance, Lionel and I saw the ISS go by, preceded by a Progress, which had got ahead of it after an aborted manoeuvre. The four years between those two flyovers in formation had been so packed with encounters and experience that I had to ask myself if I was still the same person.

From Star City we travelled to Japan, where Terry and I were trained for a few days on the Japanese experiments planned for our mission. Our back-up, Kimiya Yui, former F-15 pilot for the Japanese Self-Defence Forces, trained with us and took it upon himself to look after us while we were in Japan. He would be relieving us on the Space Station along with Kjell Lindgren, a former NASA flight surgeon and emergency medicine physician. It was while we were in Japan that I received tragic news one evening: a mid-air collision of two Italian Air Force Tornado combat jets. I was seized by a terrible foreboding and immediately called the mobile of my very dear friend Mariangela, a Tornado pilot who'd been in my class at the Air Force Academy. She didn't pick up, and I couldn't get hold of anyone I knew at her Wing either. After a miserable night, I got the news from a mutual friend: Mariangela was dead, along with Alessandro, the pilot of the other jet, and Piero Paolo and Giuseppe, the two navigators.

It was devastating news. Mariangela and I had shared a room for four years. Although very young, she'd always been a model of composure and dignity. Cheerful, strong, brilliant, and with the discipline to exploit her many talents, in a parallel world – one that was plausible and not all that different from this one – she might have been in my position, and I in hers. Her death at the age of thirty-two tormented me. I left it to Terry – who'd had his own share of such experiences – to explain my air of detachment to our Japanese instructors, and the reason I was on the verge of tears. It took me some time to get over the grief. As so often happens, I finally found peace thanks to those who were suffering the most. Two weeks after the accident, I met Mariangela's mother and her boyfriend in the funeral home, consumed with grief but calm, strong, and eager to console the rest of us. Partly because of their words, and partly due to the symbolic power of the funeral, I finally allowed myself to stop mourning and look ahead.

I left for Houston with renewed determination.

25.

> I am a part of all that I have met;
> Yet all experience is an arch wherethrough
> Gleams that untravelled world whose margin fades
> For ever and for ever when I move.
>   Alfred Tennyson, 'Ulysses'

Flight from Frankfurt to Moscow, 8 October 2014

The seatbelt light is off. For once, maybe the first time for several years, I don't take out my computer to work. I don't look out the window either – it's dark already. Nor do I start trying to talk to my neighbour: sooner or later she would ask me why I am going to Moscow, and I'd have to make something up, as always. I rarely tell the truth. As soon as the word 'astronaut' comes out of my mouth, the shockwave wipes out all possibility of having a normal, equal conversation between two human beings who enjoy learning from each other. So I just sit here looking into the emptiness and catching my breath. I've spent weeks whirling around, navigating the rapids without a break. At last the river-bed is wider, and the water is flowing peacefully. I can let myself go and do nothing.

September was frenetic in Houston. All the instructors request the crew for a final review as close to the launch as possible; missing it would spell disaster, you understand. There's a final virtual flight with the SAFER, a final Prep and Post, a final emergency simulation, a final chance to rescue a crewmate after a cardio-pulmonary arrest or to save the Space Station after a critical electric bus loss, and so forth.

The principal investigators of the human physiology

experiments also want the final pre-flight data as close to the launch as possible, which is how I found myself one day with more cooler bags and cases full of gear for experiments than I could manage to carry. If nothing else, some of the exams offered the chance to catch a half-hour nap. The countless MRIs are a case in point: to the great delight of the operators, who appreciate a still subject, I always fell asleep in a few minutes, despite the machine's thundering throughout the scan. There was no shortage, either, of final check-ups that were strict medical requirements for the Space Station crew. I had to undergo a bone-density scan again, for example, to get a pre-flight baseline; it will enable us to quantify the bone loss caused by six months in weightlessness. Similarly, an isokinetic test provided the baseline data for muscle strength and resistance; and to evaluate my balance, I took a vestibular assessment on a special moving platform.

During all of this pre-flight bustle, Alicia had to clear an entire day on my schedule so I could repeat a session in the vacuum chamber with the EMU suit. Paradoxically, while you're in Russia, this event is scheduled at the beginning of the EVA course and it's actually a prerequisite to begin training in the hydrolaboratory. In Houston, you get your vacuum chamber run at the peak of your training, as if to put the final seal on your preparation for space. When I first tried it, last July, the chamber had a technical problem. Everything went well until the pre-breathe started: a good four hours of waiting and breathing pure oxygen, since in the vacuum chamber you're not allowed to carry out the small movements to speed up the metabolism which are part of the shorter ISLE procedure. In an attempt to distract astronauts from their physical discomfort and help pass the time, they usually show a film on a screen conveniently positioned over a porthole. I chose *The Princess Bride*, an American cult comedy I absolutely had to see, they assured me. The film had only just got going when I became aware that my suit was getting warmer and warmer. This was strange, because I'd completed all the operations that required physical effort and I hadn't been moving at all for several minutes: I expected to need less cooling, not more. However, I rotated the knob to a lower temperature and, somewhat puzzled, I glanced through the pages on the LCD display until I found the temperatures in the cooling loop. They went down for a few minutes, and then started to creep back up. At that point, I reported the problem via radio, and the exercise was rapidly interrupted. There was a leak in the cooling system.

So I had to repeat the drill a couple of weeks ago. This time, I had to watch a reflection of *The Princess Bride*, since we were in a back-up chamber, and the porthole was behind me. But I was still able to clearly observe a playful physics demonstration that always takes place during this exercise. Some water is left out in an open container and, once the pressure has sufficiently decreased, it boils at room temperature. Due to restrictions in the back-up chamber, I had to forgo the usual second demonstration, in which two objects of widely differing weights and shapes are dropped simultaneously to prove that, in the absence of air, they'll reach the floor together. I'll have to go to the Moon one day in order to confirm this. Oh well!

For now, I'm going to Star City. In many ways, this trip is just like all the others. I've packed my bags at the last minute as usual, zipping up the suitcase while my taxi waited outside. I've taken the usual train to Frankfurt airport – by now I'm resigned to the habitual delays. I've boarded the usual flight, grabbing a Russian magazine at the end of the jetway to prove to myself that I can still read it – that my Russian hasn't shrunk to astronaut jargon. It's all going as usual, and yet this is no ordinary trip. The voyage towards space has begun, and I won't be going home again. My plane ticket is only one-way, and in my passport there's a visa for Kazakhstan. Actually, I won't need one to go to Baikonur, because the city has a unique legal status, but it would be helpful if, say, there were to be an emergency after the launch, and we were forced to land immediately on Kazakh soil. Strictly speaking, without a visa we'd be in the country illegally. I can see it now: us getting out of our little spaceship, somewhat battered and shaken from the hairy landing – and the first thing they ask for is our passports. After all, if we demand them of war refugees, what is an emergency re-entry from space?

Recently there have been more and more occasions when I've said to myself, 'This is the last time.' The last time in such-and-such a place or with such-and-such a person, or doing a certain thing. Of course, it's not the last time ever, at least I hope not. Just the last time before I go into space. As it happens, I don't spend a lot of time thinking about life after my mission. It's still hidden behind a thick curtain of fog.

Yesterday I went to EAC for the last time to say goodbye to my colleagues at a traditional get-together. Over the weekend I had been to Italy for the last time, just a few days in Rome and Milan, my schedule packed with outreach events, some of them pleasurable, some less so. I had to carve out a few hours for small Earth pleasures: a last pizza with a friend in Trastevere, a last Sunday walk in Sempione Park. Before that, I had said farewell to my life in Houston – maybe for ever in this case. Who knows? I may not have the good fortune to be assigned to a second ISS mission. But I feel optimistic, and I want to believe that it was only a goodbye. I was really touched by my dear friends Mary and Stacey, who threw a farewell party for me, inviting my friends and my brother's family who were in the city for the weekend. Mary and Stacey are both doctors, and for years their house has been a gathering spot for a small, close-knit group of female astronauts. They gave themselves the name Tank Girls, after the title of an awful post-apocalyptic comedy which in Cady's words is 'so bad it's good'. It's said to be a favourite of Pam's, who was first pilot and later commander of the Space Shuttle. Over the years, I always cherished the evenings at Mary and Stacey's house as safe havens of warmth and friendship. It was also humbling. Mary sometimes didn't get back in time for supper if she was working late in the operating theatre: a surgeon in one of Houston's prestigious paediatric hospitals, she saves the lives of newborn babies and children for a living. She would come home with her scrubs on and tell me in all seriousness what an exceptional job I had. Maybe . . .

Main Mission	Low Earth Orbit				Conference / Workshop	
Destination:	Neutral Zone				No	
Mission Detail:	Launch to Low Earth Orbit from Baikonur, Kasachstan.					
	After landing in Karaganda, transport back to Houston!					
Start of Work:				End of Work:		
Mission Type:	Normal			3rd Party		
Interview:	No			Justification:		
Type of Transport: Airplane						

Journey (Theoretical)						
From	To	Departure		Scheduled Arrival		Transport Detail
Town	Town	Date	Time	Date	Time	
Baikonur	Baikonur	11-NOV-2014	16:00	11-NOV-2014	16:01	---
Baikonur	Low Earth	24-NOV-2014	10:00	24-NOV-2014	10:01	other
Low Earth	Karaganda	28-MAY-2015	08:00	28-MAY-2015	11:00	other
Karaganda	Houston	28-MAY-2015	13:00	28-MAY-2015	23:30	NASA Plane
Planned Journey same as Theoretical						

Mission order, formally approved by my manager like
a regular business trip

As the plane lands in Moscow, applause breaks out, a liberating gesture I've only seen Russians and Italians make before. Twin peoples – I have always known that. Before I get off the plane I check and recheck to make sure I haven't forgotten anything. I have a very precious piece of hand luggage with me and I certainly don't want to lose it on the way to Moscow, considering that I'm supposed to take care of it for the entire trip to space, there and back. Inside are jewellery and photographs, but also patches, stones, small toys and children's drawings. These objects of great sentimental value have been entrusted to me by family and friends, and I'm allowed to take them with me on the Soyuz as long as all of them weigh less than 1.5 kg. Barring any unforeseen circumstances, they'll all come back to Earth with us. I also have about 150 copies of the minibook containing my selection of texts. I've called it *Untravelled World*, and on the cover is a lovely drawing by my pen-pal, Giorgia, who works in data visualization. 'Pen-pal' in an era of piecemeal communication: we've never exchanged letters, only private Twitter messages. I have thousands of copies of my booklet, since it wasn't possible to print fewer, but 500 of them are numbered. It was Kjell's three young children who took care of stamping the numbers in them a few weeks ago, or so he assured me. I suspect that he helped himself, to relieve me of the worry. Good old Kjell. The 350 numbered booklets I don't have with me were left in Houston, and if all goes to plan they'll be included in my Crew Care Packages, the psychological support kits that get put together by the families and sent aboard. With a bit of luck, it will be possible to send them back down on a Dragon.

In my hand baggage I'm also carrying the accessories I'll have with me on the Soyuz. They gave them to me in Houston, an envelope full of items I'd chosen from a list a few months ago: a kneeboard with ten sheets of white paper, pen and pencil, two lanyards to secure them with, a torch, two bulldog clips, a pack of small Velcro squares, a cheap stopwatch and some rather expensive (so they tell me) sickbags. I asked for four of them, somewhat guiltily. Our journey to the ISS is expected to be a fast one, six hours from launch to docking. But it was the same schedule for the crew who left in March, and they had to transition to the old two-day rendezvous profile after a computer bug disrupted an engine burn. The Soyuz carrying Sasha, Yelena and Butch in September was also a bit capricious: after orbital injection, only one solar panel extended, and the second one was deployed only after docking. So it's better to be prepared. Four sickbags.

———

The next five weeks in Star City were in most ways a repeat of those I'd spent in the spring as back-up crew: the final training sessions, Sundays spent studying, the visit to TsUP, some half dozen exams, the rituals prior to departing for Baikonur. Ironically, now that things were becoming serious, I wasn't experiencing those moments with the same intensity as before. I just didn't feel the same childlike enthusiasm of the first time, or even the pre-exam nerves. The easygoing camaraderie with Reid and Alex was missing. It was all a bit dulled by repetition, and even the colours soon faded. The golden autumn was over, and the first snows had arrived in mid-October. By this point I felt caught up in the final countdown.

However, life still had a surprise in store for me before I took off for space. On 28 October, while I was sleeping peacefully after an evening going over Soyuz procedures with Kimiya, the uber-prepared back-up-crew flight engineer, an Antares rocket exploded on the other side of the world a few seconds after lift-off. Fortunately, there were no victims. On board was a cargo vehicle, Cygnus, with more than two tonnes of provisions for the Space Station – all lost. I heard about it the next morning, when I saw all the emails in my inbox from NASA, ESA and ASI with the first bits of news. It had happened only a few hours before, but everyone had promptly gone to work to evaluate the impact and begin sketching a plan for replacing what had been destroyed.

Initially, I absorbed the news without making much of it and even somewhat distractedly, because our final exams on the Russian segment of the ISS and on Soyuz were scheduled for the next two days. I was determined to stay focused on this last duty, and ignore the issue with the Cygnus. In point of fact, the actual event seemed unsurprising. The American commercial cargo vehicles were NASA's attempt to encourage the development of the national space industry, and this approach carried with it the conscious acceptance of some risk. I trusted that the supply chain of the Space Station had been managed in such a way as to absorb the cargo loss with limited consequences. Only later did I learn that one of those consequences, however irrelevant to life on the Space Station overall, would mean a great disappointment for me.

The exam on the Russian segment was an easy task for me and Terry, as it was the first time. The memorable part of the day occurred before we began, during the usual pre-exam interviews. We still talk about it, and it never fails to make us crack up. Here is what happened: a journalist asked me in all apparent seriousness how it felt to be going into space with two such attractive men, whose strong shoulders I could lean on. From that moment on, Terry had only to glance suggestively at one of his shoulders to get us all laughing.

The next day we showed up again to face the committee for the final Soyuz exams. After three and a half years of training, the last – the actual *last* – day had finally arrived. After the usual grouchy but good-natured quips from Valery Korzun, oscillating, as always, between scolding and encouragement, Anton chose one of the envelopes with the exam scenarios. Kjell, Kimiya and Oleg, our back-up crew, had drawn the envelope with the fire the day before, so we knew for sure that this time we weren't about to burn up. And in fact, Anton chose a really lucky envelope: there were five failures, as always, but all of them were easily managed. So we finished the exam session with little effort, but with top marks.

The festivities were memorable. They began as usual with refreshments in a nearby meeting room, where Terry, Anton and I took our places, along with the back-up crew, at the head of a long table made of desks pushed together. It was covered with plastic plates piled high with fruit, tomatoes, cucumbers and cold cuts. There was water, fruit juice, and lots and lots of vodka. After a few hours and countless toasts, we moved on to Cottage 3, where the festivities went on into the night on a variety of themes: celebrating our exams, typical Friday-night cheer and Halloween for NASA staff and their families. On the stairs there was a pumpkin carved with the numbers 42 and 43 – the work of Sasha, our manual-docking instructor. We were as happy as kids running around an amusement park.

That weekend, my mind free – though perhaps not entirely lucid! – I began reading emails about the exploded Cygnus which had accumulated over the past few days. The immediate impact was felt on the Space Station's current activities, which would need to be replanned. I imagined dozens of schedulers working feverishly to reshuffle the timeline of the next weeks, taking interdependence, conflicts, readiness of the procedures, the availability of ground support and much more into consideration.

It was less urgent but just as important to evaluate how and when to replace the items that had been destroyed: experiment hardware, spare parts, EVA equipment. Luckily none of us had lost anything truly personal. My crew-preference bag, which I had filled mostly with tracksuits and soft, warm fleeces, was already on the ISS: Alex even sent a photo to reassure me. After some uncertainty about three of the containers, I was also told that all my bonus food was on board too, and half my clothes. The other half had burned up in the Cygnus, but there was still time to replace them. The next launches would be our Soyuz, with its small cargo volume, and the Dragon of the SpaceX-5 mission, the latter being the first real opportunity for any important resupply. It was planned for December, and there were enough provisions on board to last until then, with one notable exception: tissues. I was confident we could live with that.

There was another email chain about my personal EVA kit. In addition to gloves, every astronaut has a green net bag sent up, full of personal items such as underwear, padded glove inserts, mole-skin patches, and most importantly, the LCVG cooling garment. My bag had been on the Cygnus. This was not good news, particularly just when a third EVA during our expedition was being mooted, and I had got my hopes up. But I wasn't all that worried. I'd got used to the idea that in order to guarantee the necessary redundancy, it was essential that all three of us should be able to perform an EVA, and that was one of the reasons why I had applied myself so vigorously to the training. If I were not ready to perform a spacewalk, it would need only a health concern regarding Terry or Butch for us to lose EVA capability with the EMU suit. Moreover, the emails arriving from Houston – some of them official, others informal – all sent the same message: the team were already working on a replacement plan. A few days later, more detailed, yet still reassuring news arrived: they'd taken my gloves off the Soyuz because they had to free up space for more urgent things, but they'd send them up in December on Dragon, along with the replacement for my LCVG. There didn't seem to be any hitches, and I was convinced – perhaps a bit naively – that everything was taken care of.

After our final exams, there were two days of holiday in Russia in celebration of national unity. Added to the weekend, this meant that we had a good ten days free before our departure for Baikonur. Anton invited me to spend a couple of days with him, Oleg and their respective families in a sanatorium, a large, half-empty hotel, something between a clinic and a health spa, and built in the rather drab Soviet style. Cosmonauts love to go there for a break. We stopped on the way to visit the fabulous monastery of Sergiyev Posad, where our guide was Father Ioav. He lives in the monastery but sees to the little church and the faithful from Star City. In his black robes and with his long beard, he was one of those religious people so brimming with love and gentleness that you want to share their faith, even if only to make them happy. I'd often seen him at the Profilaktorium, and he loved telling me about his trips to the Italian city of Bari. He'd been there more than once on pilgrimage to the Basilica of S. Nicola, a place of worship for both Catholics and Eastern Orthodox believers. He offers each crew the chance to pray with him in the monastery before a launch.

And I think everyone accepts, whether they believe or not.

26.

Baikonur, 23 November 2014

The countdown is over. Today I woke up on Earth and I'll fall asleep in space.

My programme for the day requires a time-out: five hours of sleep until 5.00 p.m., before the meticulous pre-launch choreography begins. I'd have been very glad for some rest, since it'll be more or less midday tomorrow before I can close my eyes again on the Space Station, yet despite the small dose of sleeping medicine Brigitte prescribed, I woke up well before time. Maybe it's just my usual strict biological clock that rebels at having to take an afternoon nap, or maybe I'm more nervous than I care to admit. At any rate, I'm awake, under the covers, with my feet elevated thanks to a mattress raised a few centimetres at one end, a little trick that's thought to help your system adapt to weightlessness. Another is to spend the odd half hour on a table tilted at a 45-degree angle, with your head down. I did that a few times during the quarantine and I took a photo showing the face I may have in space: swollen, with eyes narrowed due to the fluid that's drained from my body into my head.

I adore a good lie-in. It's one of my great weaknesses. Unless I have jet lag, I fall asleep instantly at night, but I always linger under the covers in the morning. I wonder if it'll be the same in space, and if I'll still sleep well without the sensation of my body sinking into a mattress. I've heard that many astronauts don't enjoy sleep floating, so they use bungees to anchor themselves to a wall in order to feel some pressure.

My last nap on Earth was strange, as if part of me had

remained awake to look on while the other part slept. These last weeks have been like that, to be honest: there I was, taking part in the events, lessons and rituals, but I was also elsewhere, looking in from the outside, as if scrolling through frames of a film. The closer it gets to the launch, the more detached I feel. Maybe a sort of unconscious wisdom is slowly shaping my spirit to prepare me for what lies ahead: letting myself be moved by a launch as I have in the past, as a spectator, is a luxury I can't afford this time. For those actually in the rocket, a launch must be reduced to a matter of procedure and technology. You can harbour both dazzled amazement and clear detachment in your heart, but the most appropriate attitude must prevail when the time comes.

In my half-sleeping, half-waking state, I can't decide whether to get up or try to sleep a bit longer. I look back on our experience in Baikonur as prime crew and imagine it depicted as a cycle of frescos, like those in a church that tell the lives of saints, always showing the same scenes, more or less: birth, meetings, miracles, martyrdom. We're not saints, we don't perform miracles and we're certainly not aiming for martyrdom, but I can imagine a few fitting scenes. There we are, leaving Star City on the bus for Chkalovsky airport, our hands pressed against the window, overlapping those of our loved ones outside. In a corner – and this is the beauty of frescos – is a parallel scene: Oleg, Kimiya and Kjell turning back after having seen us off. One of the planes, in fact, has broken down, and they will only be able to get to Baikonur the following day. In the next scene, one of us is crawling through the hatch of the orbital module in a blue suit and a white chef's hat, while technicians bustle about in surgical masks. In another frame we're putting on our Sokol suits; still another shows us in the descent module with all our belts fastened, checking our work envelope and ability to reach and operate the systems. It's our first fit check, the first time we see our spaceship again after having left it all those months ago in Moscow.

There's a further fresco dedicated to the quarantine: here we are in class with Dima, going over procedures; there we're training for manual docking with a portable simulator sitting on the table; in another frame we are sitting down next to each other as, assembly-line style, we each sign one of the hundreds, perhaps thousands, of official crew photos piled on tables and chairs. In a scene that feels almost domestic, Dima and I are checking the weight of my personal objects, individually numbered and packed in small plastic bags.

It's strange how wrapped up we get in details and concerns of surprising banality the closer we get to the launch. Where will I find my clothes on board? A short time ago, Bernadette sent me an Excel summary sheet, but where did I put it? What can I take for nausea, if I succumb to the infamous space sickness? Should I take a preventative medicine before we launch? What would Brigitte recommend? And what do Reid and Alex think about it? They're in space, but after all only an email away . . . How should I take phone numbers and passwords with me? I've got them all on one page, which Lionel will send me by email as soon as our on-board account is active upon my arrival. But I have a paper copy too, and I've given it to Dima to put in one of our procedure checklists. It also has my credit card details, but that card was blocked a few days ago because of a suspicious attempt to use it on the other side of the world. So no purchases from the ISS . . . I'll get over it.

The second and final fit check of our Soyuz took place a few days ago. The spaceship was already shrouded in the fairing, and our visit, complete with signatures on the forms, was now all that was missing before the Soyuz could be mounted on top of the rocket. We climbed up as usual to the third level of the yellow scaffolding that wrapped around the Soyuz and crawled in through the small opening in the fairing that corresponds to the Soyuz access hatch. Inside the orbital module, there was hardly any space to move because of all the cargo that was carefully arranged in the cabin, most of it packaged in clear anti-static bags. I recognized my gym shoes in one of them, and also my cycling shoes, the only footwear I'll need in space. There was no point in looking for my EVA gloves, though. I knew very well that they wouldn't be there, that they'd been scratched from the cargo list after the loss of the Cygnus to make room for more urgent supplies. No harm done as long as they were coming on Dragon in a few weeks as previously planned. But it won't happen. They'll be staying on Earth. I became aware of this during the peaceful routine of the first week of quarantine, and the news coated me with a sticky bitterness that I'm still trying to shake off completely. Due to an intercontinental misunderstanding over who would tell me about it, and when, I found out in a cold and brutal way: I opened an email attachment with a table showing the new cargo manifest for the next launches, and in the Dragon column my gloves and the sack containing my personal EVA equipment were crossed out and highlighted in red, with a brief note in the margin explaining that, due to lack of space, the ISS programme management had decided not to launch them, and to accept the risk of having only two crew members who could perform an EVA with the EMU suit. I read it several times, but there was not a shred of doubt, not the least uncertainty. The message was clear: my EVA equipment would not be on board. I felt dreadful, a mixture of indignation and self-pity. How could I have been so mistaken all this time, believing that redundancy of EVA capability was essential, to the point where it was unthinkable to consider being part of an ISS crew if you hadn't demonstrated during training that you could perform a spacewalk? Had all the late evenings and entire weekends I'd spent preparing for NBL runs been misplaced effort, the absurd result of some misunderstanding? These were all pointless questions at that stage, but they went round and round in my head. I couldn't stop them any more than I could have stopped the waves in the sea from crashing to the shore during a storm.

Obviously there's nothing for me to get angry about. No one has done me any wrong, and neither the world nor the ISS revolves around my personal ambitions. My dream of making a spacewalk has come up against the reality of some unlucky circumstances – full stop. The disappointment burns and will continue to burn for a long time, but you can't always have good luck, and I've had a great deal in my life. Ironically, for a couple of weeks it seemed like the loss of Cygnus might even have a silver lining. The accident, and the resulting no-show for many spare parts and experiment hardware, had eased the shortage of astronaut time that afflicted our mission, and it would have been easier, now, to put on the timeline the assembly of a medium-sized suit, news that had been more welcome than ever now that a third EVA for Expedition 42 seemed ever more likely. Yet just when it seemed that a window of opportunity had opened, all hope flew out of it.

Maybe I was wrong to want an EVA so badly, when so many people had warned me that I should not get too carried away about it, since even at the best of times it's uncertain and liable to unforeseen changes of plan. In retrospect, it's easy to regret not being more detached about it. The truth is that if things had gone differently, I would have been pleased with my own tenacity and praised for my grit. I don't want to fall into the trap of judging the value of an endeavour according to whether it's successful or not. I don't want to accept that the success of a venture decides, after the event, whether pursuing an improbable goal was determination or naivety, tenacity or obsession. I reminded myself of the words I love to repeat to young boys and girls whenever I meet them: if you have an ambition, take motivation from that and apply yourself to the utmost. Choose the most difficult path, the one that allows you to grow. It's important to have a dream for the journey, not for the finish line. As they say in Houston, EVA training builds character. It's made me stronger and more aware of my limits, but also of my strengths. It's taught me how to ration my energy, streamline my movements and actions, to perform every movement in a considered, conscious way. I've learned to manage my frustration, change orientation – literally – to identify a different approach to a difficult task, and to find the right balance between carrying on and looking for an alternative solution. What's about to begin won't, I hope, be my last mission, and maybe next time I'll be able to make a spacewalk. For now, thanks to my EVA training, I'm a little better for it.

I'll never get back to sleep; I have to accept that. I get up and reach for my computer, struck by how nearly empty the place is now, like an unoccupied hotel room. My luggage is waiting by the door. I've packed a couple of suitcases and various bags, meticulously organized according to destination: some things can just go back home, but I'll need others in Houston on my return, since after landing, Terry and I will be taken there directly. I've labelled clearly, or at least I hope I have, the things that will remain in Star City and be transferred to the two bags which will be taken to the nominal landing site and the ballistic one, respectively, on the day of our re-entry. According to instructions, in each bag I've placed sunglasses, gym shoes, a change of underwear, a blue flight suit for official welcoming ceremonies, sportswear for the long flight to Houston, toothbrush and toothpaste, deodorant, and a bottle of shampoo for my first real shower after six months. There are also some snacks: a few weeks ago, I found myself in a Moscow supermarket, thinking about what I'd want to eat when I got back from space. I placed my bets on Mediterranean flavours: crackers with rosemary and *taralli* with olive oil.

I go back to bed, arranging the cushions so I can lean on them comfortably, and open up my laptop. The good wishes keep coming by email, many of them from people I haven't been in contact with for years. They make me happy, but I'm not responding to them any more. Yesterday evening, I set up an auto-response: 'I'll be off the planet for a while, back in May 2015. Unfortunately I won't read your message.' While I'm on the ISS I'll have an ad hoc email account, but I'll only receive messages from authorized senders. One of the many things keeping me busy over the past few weeks was putting that list together and including key people from the team I work with, along with family and friends.

Some of them are here in Baikonur now – fifteen in all, including Lionel, my parents and my brother: that's how many Roscosmos allows each astronaut to invite, though the cost of getting here is prohibitive for most. My guests arrived in Kazakhstan the other day after a few days in Moscow and Star City, where they were entrusted to the care of my EAC colleagues Romain and Manuela, and Lionel, who never tires of organizing visits or booking restaurants for the little group. I couldn't have direct contact with them because of quarantine restrictions, but we had a couple of half-hour sessions to talk through a glass partition. I sat on one side and they on the other, in conference-room chairs, passing a microphone around in order to talk to me. There's no use pretending that the situation wasn't a little awkward, but on the other hand it was light-hearted, and there was a nice bond between them, despite the fact that many of them had never met one another before arriving in Moscow. Lionel busied himself preparing a little gift box for everyone including a towel since, as *The Hitchhiker's Guide* points out, 'A towel . . . is about the most massively useful thing an interstellar hitchhiker can have.' If this launch goes belly-up, it won't be for want of towels.

My brother, my parents and of course Lionel were allowed to meet with me without the glass wall in the way, though they had to submit to a daily visit with Doctor Savin. We were repeatedly reminded that we should avoid all physical contact, and we all heeded the recommendations, within reasonable limits. We went for a walk along Cosmonauts Alley and I took them to see the little tree I planted a few days ago. We then went to the recreation room for a chat, before my brother and my parents left me alone with Lionel for a few moments. Lionel came back for a final time last night to watch the traditional screening of *White Sun of the Desert*, a classic film from the Soviet era which is shown to crews on the eve of every launch. We watched part of it, and then we took some time to say goodbye in private, giving in to a rather emotional parting.

Still on the bed, I write a few spontaneous words of thanks on the last Earth-based page of my logbook, a near-daily blog I started more than a year ago, 500 days before the launch. It's been a long journey to get where I am today, and it's impossible to remember everyone who has helped me, encouraged me, looked on with kindness or shown patience with my mistakes and excesses, which are exacerbated by my somewhat tough character. And when that character earned me someone's hostility, it's likely that even that was for the good. Life is made up of complex and unpredictable relationships, and my being here today is the result of all that's come before, no matter whether something seemed to me like a good or a bad thing at the time. Even those who've tried to obstruct my progress have become stepping stones on my path. I'll take them with me into space, along with every other person I've met, everything I've learned and every experience I've lived through.

It's time now. Someone's knocking at the door. It's Antonio, EAC's rep for crew support; he's gone through quarantine with me. Prompt as always, he's here to take instructions about my baggage, and then he'll leave me alone with Brigitte for a last medical check-up. Olga, the Russian doctor, joins us after a few minutes for what is euphemistically termed 'special medical procedure' on our schedule for the day: an enema. It's not obligatory, but it's heartily recommended, and I don't know of any astronaut who failed to accept this offer willingly, in hopes that they wouldn't have to do a Number Two in the Soyuz toilet. Olga is an expert and carries out the unpleasant procedure quickly and efficiently. Then I withdraw to the bathroom, where Doctor Savin has left a bowl of alcohol and a few large towelettes for me to use to disinfect the skin on my entire body. But first I linger in the shower, surrendering to the pleasant sensation of hot water running through my hair. It won't happen again for a long time.

When I've finished my cleansing procedures, I put on the Sokol undergarments, white cotton leggings and a long-sleeved vest. I pull a t-shirt and sports trousers over those and then join Terry and Anton in the small dining room for our final meal. It's been dark outside for several hours, but our day is only just beginning. There are eight hours to go before the launch. We were asked in advance what we wanted to eat for every meal while we were in quarantine, but not this time: the kitchen has clear instructions, and I don't know if it's down to tradition or prudent choice. We are served two similar dishes, a stuffed focaccia and some sort of thick, dense quiche, which I remember eating before Reid, Alex and Maksim's launch. It's not a memorable meal, but that's OK. I'm fine with eating whatever Russian wisdom considers least likely to show up again in a few hours, should I discover that I am one of those astronauts who suffer space sickness. We eat quickly, saying little, and return to our rooms for a short while. At 19.40 local time, we leave our quarters, never to return.

The air is crisp, but it's not an unpleasant cold, just nippy enough to bring my thoughts back to the present – they're prone to racing ahead. In a couple of minutes we arrive at the main building and go up to the second floor, to the wing the crew used to stay in until a few years ago. We raise a toast in what was then the commander's room. Luckily, it's the last toast. I've run out of ideas, and maybe patience, too. Terry and Anton's wives are here, and Lionel too, along with reps from NASA, ESA and Roscosmos. We toast with fruit juice, but the back-up crew drinks sparkling wine as if symbolically to underline the fact that they are definitely not the ones going into space today. Waiting for us in the corridor is a small crowd composed of our families, reps from various agencies and a few photographers, ready to immortalize the umpteenth ritual gesture: signing our names on the door. A little way away, an orthodox priest wearing yellow vestments is waiting for us in a small antechamber, ready to give us his blessing. As we pause before him, he dips a large sprinkler in a golden cup, enthusiastically waving it towards me, the first one in the queue.

And so it's with my face and hair dripping with holy water and bolstered by divine protection that I prepare to leave quarantine definitively and proceed to the most important night of my life. The loudspeakers are blasting out a remix of the famous Russian rock song 'Trava u doma', 'Grass by the Home', considered the official cosmonauts' hymn. Anton, Terry and I stride purposefully down the stairs to this soundtrack, lined up beside each other, and we walk out of the main entrance. Shouts of encouragement are coming from the courtyard: we can't see them since only the walkway is illuminated, but behind the barriers, our families and friends are rooting for us. We wave to them, smiling blindly into the dark as we walk through cold air thrumming with music and frenzied yelling. We climb into the bus and put our hands to the windows, meeting those of our loved ones through the glass. Minutes later, we're on the road that crosses the steppe to the cosmodrome. It's a deserted road, apart from a few police cars here and there, the policemen standing beside their vehicles to offer us a military salute as we go by.

About forty minutes later we arrive at the now familiar Building 254.

As always, at the entrance we place our feet in a machine that wraps them in plastic shoe covers. From there, we go into the small crew room where we've waited during the breaks on the long fit check days. There are a couple of faux-leather sofas and a small television on a stand, tuned as always to a music-video channel. In the room next door, there are tea and biscuits. I eat one or two while we wait. It seems like any other day.

They call for me about twenty minutes later. As the flight engineer, I must put on the Sokol suit first. I go to the loo – one more 'last time' – pausing to appreciate the simplicity of an Earth toilet. Who knows how it'll go in space. It's things like this that worry me. If the rocket should explode, well, there's not a lot I can do, but I want to make sure I can take a pee in space without surrendering my dignity. The first time will almost certainly be in my nappy. It seems that Russian cosmonauts don't wear them, perhaps finding them embarrassing, but it's popular with the rest of the community. We all remember the flight of the first American astronaut, Alan Shepard, who was forced to empty his bladder in his suit as he waited for the launch, short-circuiting the medical sensors.

I remove my shoes and my clothes, apart from my underwear, and put everything in a bag, which I give to Brigitte along with my documents, telephone and wallet, to be kept in a safe in Star City until I get back. I've now left everything behind, apart from the few things I've got on. Technicians are waiting in the room next door to help me put on the suit. The Sokol is a single piece with two large zippers that run from stomach to shoulder.

You put your legs in first, and then your torso, through the front opening; you have to duck your head to get it through the neck ring. You feel trapped the first few times you do it, but you get used to it. Once your head, arms and hands are inside the suit, you are left with an untidy bundle of excess material from the lining around your stomach; it's this internal membrane that keeps the Sokol pressurized. With gestures that have an almost domestic character and seem in harmony with the warm light, laminate flooring and large flowered carpet in this room, I start folding the lining neatly, tying it with thick elastic laces as I've been taught to do. I could keep going with my eyes closed, closing straps, hooks and zippers, a sequence I know by heart and one that keeps my hands busy and my mind focused on the present, on a simple and well-defined task. Putting on my suit is the first thing I do tonight as the Soyuz flight engineer. It's starting. I feel calm and ready.

When all three of us are dressed, we proceed to the leak check, which for some ludicrous reason takes place by tradition in the presence of journalists and our families, all of them lined up behind a glass partition. As we complete the check, one by one, we're invited to take a seat at the table with the glass partition, and we take turns using the microphone to speak to our loved ones on the other side. The crews don't like this practice very much, and I feel awkward about it too. I've already had a chance to say goodbye to Lionel, my parents and my brother. What else can we say to each other in front of these journalists and all the television cameras? We put a brave face on it, chatting about this and that, and after what seems an endless amount of time, we hear the liberating announcement that it's time to go and meet our rocket.

There's almost an hour's journey ahead of us because we leave from Ramp 31, about 40 kilometres away, since the nearby Ramp 1, which has been used since Yuri Gagarin's launch, is closed for maintenance. We keep the lights off in the bus and make a few quips, but it's mostly silent. For part of the journey I decide to sit in front beside the driver, looking out at the road and quietly taking leave of this planet. A police helicopter leads the way, sweeping the area in front of us with its searchlight. The rhythmic effect is hypnotic and I feel like I could sleep. There are only three hours to go before the launch; shouldn't I be throbbing with adrenaline and tension? I've never thought much about the risks, but I did expect to feel a little apprehensive when we got this close. Of course, the Soyuz is well known to be a rocket of proven reliability. Maybe I would have been more nervous on the Space Shuttle, which has been involved in two catastrophic accidents. I wonder, too, if it's more difficult for Anton and Terry, both of them fathers with young children. If that's the case, they're not showing it, but we have never really talked about it. I don't think anyone ever talks about it. I sit down on my large, comfortable seat and go to sleep.

Olga, the Russian doctor, wakes me up a few minutes before we get to the ramp and hands me a bottle of water and two tablets from Brigitte.

One of them is a preventative to block the symptoms of space sickness, and the other is an ordinary painkiller; we'll be stuck in the foetal position for hours. A little later, the bus stops to allow us to complete the final ritual gesture: urinating on the wheel, as Yuri Gagarin did in 1961 and, it seems, every male cosmonaut or astronaut who has since left for space from Baikonur. In view of the difficulty of getting out of the suit, I don't believe any woman has ever done it. I know that some of my colleagues brought pee in a small bottle so they could throw it over the wheel, but I've never felt tempted to curry favour from the fates this way. I suspect that the survival of the tradition has more to do with practical reasons than superstition. As we admire the waiting rocket, shining gloriously in the night, I can't help but envy Terry and Anton their relief.

27.

Baikonur Cosmodrome, 24 November 2014

The wind turns suddenly, and the white vapour dancing around the scaffolding on the launchpad wraps us in its icy embrace, hiding us from the view of the group of men who accompanied us to this point. It's a final, kindly gesture from the Earth, which seems to want to ease our separation by means of an impersonal, emphatic intervention. We're not the ones who want to put an end to this farewell ritual, on the stairway with the yellow hand-rail at the foot of the rocket shining in the night, and our hands, in thin white glove-liners, waving goodbye as our lips smile genuinely, but are also genuinely tired. We're not the ones who want to evade the last photos, the last sound of our names called into the air over the muffled hum of the launchpad. Please forgive us. We're not the ones burning with desire to turn around and keep climbing, to become one with the rocket. It's the rocket that calls out to us, breathes over us this white, cold fog, that swallows us in a spectral light. End of the scene, end of the film. The end of exams, rituals, goodbyes, interviews, preparations, photos, celebrations, toasts. It's all been splendid, but now, forgive us if we're in a hurry. There's a rocket here, ready to take us into space.

I catch sight of it for a moment, so close I can almost touch it, clad in ice because of the extremely cold temperature in the liquid oxygen tanks. Frost, I think. Our rocket is covered in frost, like the windows I used to admire, fascinated, on winter mornings in my childhood. A few final bumbling steps in our Sokol suits, so hostile to the standing position, and we find ourselves in a small wobbly lift that starts to move slowly, clanking as it goes.

Our escort, a young technician, agrees to take a photo of us on his mobile. If there should be a fatal accident during the launch, this will be the last remaining photo of us. I like it: we look radiant and bonded.

When the lift stops, the technician opens the door and leads us into a bare, poky room. Our Soyuz is waiting on this level, crouched over 300 tonnes of kerosene and liquid oxygen and enclosed in its protective fairing. As the flight engineer, I prepare to enter first. I detach the suit from my portable fan, a grey metal box that I will not need any more, because I'll soon be connected to the Soyuz ventilation system. My boots are also useless now. I take them off, and crawl in through the opening in the fairing and through the access hatch of the orbital module.

Everything is exactly as I expected. The cargo packs are wrapped and attached just as we left them four days ago after our final fit check. Fit check, we say. Like trying on clothes at the tailor. In effect, our spaceship is so small that it doesn't take much to imagine putting it on like an item of clothing. It also has a characteristic and unmistakable odour that immediately elicits for me – with the immediacy that only olfactory stimuli can evoke – a sense of calm familiarity. However exceptional this voyage I'm about to undertake, however undeniable the risk of a sudden and violent death tonight, the truth is: I feel at home. There's no other place in the world I'd rather be, nothing I know as well as this spacecraft, no situation in which I'd feel so completely and unconditionally that it was the right place for me. It's time to begin this adventure.

I slowly let myself into the descent module, taking care not to damage the open hatch since I have to slide beside it, first down, towards the middle seat, and then sideways towards my seat on the left. Before doing anything else, I attach myself to the Soyuz using the appendages on my suit: the oxygen and ventilation tubes, and the audio and medical telemetry cables. I'm already wearing the *shlemofon* the cloth headgear with headphones and microphone.

After connecting the cables, I slide into my place, shifting my back so that it fits against the seat liner. Huddled in the usual foetal position, I gladly allow the technician to help me strap in, saving myself the contortions I'd otherwise have to make in order to reach the countless hooks and straps on my own. For now, I'm leaving them fairly loose; there's no reason to restrict my freedom of movement too much. About forty minutes before launch, we'll receive precise instructions to tighten the straps as much as we can, taking care first to let all the air out of our lungs. Only after we confirm this will they activate the launch escape system, the complex of thrusters and pyrotechnic devices that will, if necessary, take us to safety with extreme acceleration. If our straps weren't tightly fastened, we would never survive the violence of an emergency escape unharmed.

Back in the orbital module, the technician passes me my procedure checklists. The one I need now is called Quick Rendezvous, and it contains all the normal procedures right up to docking on the Space Station six hours after launch or, as we commonly say, after four orbits. Quick Rendezvous contains roughly a hundred pages which are slightly smaller than A4 and is ring-bound, with colourful hard covers. Over the past few days Dima has applied adhesive rings to reinforce every hole so that the pages won't come loose when we turn them with our gloves on.

After handing me the checklists, the technician passes me the extendable baton I need to reach the control buttons higher up on the panel. It's usually only the commander who uses the baton from the central seat, which is set back from the others, but I'm small and didn't want to take any risks. At my request, the Soyuz TMA-15M spacecraft must have two batons on board: it's undoubtedly written down on some form, signed and stamped.

It's Terry's turn to get into the orbital module. While the technician helps him to get strapped in, I establish contact with the bunker, the location for the launch control centre. I switch on the communication equipment and get ready to make the first radio call of this long night. We, the Astrey crew, are about to make ourselves heard from that undefined location between space and Earth that is a rocket, full of fuel and ready to launch.

'16-3rd, this is Astrey 2, how do you hear me?'

16-3rd, or 'Sixteen-third', is the mysterious call sign of the post assigned to speak to the crew from the bunker. The reasoning behind this name has always eluded me. Or maybe I never raised this question with adequate persistence, as with thousands of other ones I might have asked on the long road to this launch had there been the time. I'm an astronaut. My job is not to satisfy my curiosity, but to sift through the mountains of information for what I need to bring my mission about. It's an operational perspective, not an academic one.

Anton, too, has finally dropped into the descent module and closed the hatch behind him. Terry and I get to work, helping him as well as we can to strap in, and then 16-3rd informs us that our vital signs are now being recorded. We need to stay still for about ten minutes. In this short time-out, the powerful presence of the rocket makes itself felt, this beast, still motionless, but slowly coming alive. I close my eyes, focusing on the noises coming from the depths of the launchpad, which break into the monotonous hum of the suit's ventilation system. Rustling, vibrations, metallic sounds, whistles, puffing . . . This mighty metal animal, which has swallowed us, is preparing its limbs, made of pumps, pipes, valves, nozzles and actuators, so that it will be ready, in less than two hours, to release 500 tonnes of thrust in a burning inferno that will light up the sky over the Baikonur steppe.

The bunker informs us that they're checking the images of the two internal video cameras: they can see Anton and me on the first one and Terry on the second. Anton takes the opportunity to ask if 'Olaf', a little white doll we've suspended from a cord as a 'weightlessness indicator', might not interfere with the shots. I think what Anton really wants to know is whether Olaf can easily be seen on camera. As the last rocket stage is shut down, Olaf will begin to float, indicating that we're in orbit. Anton must surely be thinking about Kira, his youngest child, who's watching the launch here in Baikonur along with the rest of our families. It was she who chose Olaf.

Once our vitals have been taken and recorded, we begin the pre-launch checks, working systematically. Anton reads the procedures out loud, line by line, and together we verify the position of the switches, physical or virtual, the parameter values, the status of the warning lights and alarms. As commander, Anton is responsible for managing the operations and he makes most of the communications with the bunker. My responsibility as flight engineer is first of all to support him, but also to pay particular attention to the status of the systems. Going beyond what is explicitly written in the procedures, I regularly check the parameters so I'll recognize any variations from the norm well before an alarm goes off. There's actually no alarm for some fairly serious problems, such as a leak in one of the helium tanks that pressurize the Soyuz engines, and the only way of discovering a malfunction promptly is through constant checking. For other issues, there are red warning lights on the control panel. One of these is particularly worrying: launcher failure. One doesn't pilot a rocket. As far as the crew is concerned, it either functions or it doesn't. It takes you into orbit, or it tries to kill you. There's no third option.

We're getting through the operations in good time. Among the many verifications, one of the most important is the leak check of the hatch between us and the orbital module. If there were to be an emergency and we were forced to make a re-entry during the ascent or right after orbital injection, the three parts of our Soyuz would separate, and this hatch would be the only thing remaining between us and the vacuum of space, and the violence and heat during the re-entry. It's therefore essential to ensure that it's hermetically sealed. We also pay great attention to inputting the data of the first two manoeuvres, which will be performed automatically once we arrive in orbit. We're telling the computer when to ignite the engine, how long to leave it on and what attitude to do the burn in: a sizeable mistake in these data could compromise our mission, as it would mean wasting fuel on a wrong orbital correction.

When we finally complete a successful leak check on our Sokol suits, there's nothing more to do but wait. I feel calm, almost euphoric. For our friends and families outside, it's past two in the morning, but in Moscow it's only 23.15, and in the few cubic metres of space we still claim on Earth, those marked out by the metal shell of our Soyuz, it's Moscow time. The launch will take place one minute past midnight.

Our vitals are logged once more, and then 16-3rd offers to broadcast the songs we've chosen for this waiting period. None of us knows what the others have chosen, but Anton's choice of 'The Final Countdown' is no surprise. The next song, 'Oh, What a Night', is my choice and elicits Terry's enthusiastic approval. Really, what a night, what a perfect night! I couldn't have imagined anything better, and I couldn't be feeling more at peace with myself than I do now.

A country song is up next. I don't know it, but I smile at the words: 'Come up here, take the wheel, I'm gonna let you drive for the first time.' Terry chose it for me. I'm the only rookie in the crew, the only one on her first mission. Around four minutes past midnight, about 200 seconds into our ascent, I'll surpass an altitude of 100 kilometres for the first time; that's where space begins, by convention. For the past five years, I've belonged to the European astronaut corps, but only in that moment will I truly become an astronaut.

First, however, I have something very important to do: pee. I don't feel the need to go yet, but we're talking about preventative measures. Too many veteran astronauts have told me how difficult it is, at first, to empty your bladder in weightlessness, especially if you have to do it in your nappy. As flight engineer, I cannot leave my seat until we get to the ISS, so it's better to take advantage of these last few minutes on the launchpad, and leave Earth as one should – with an empty bladder.

Only twenty minutes to go, and I'm checking our parameters for the umpteenth time when I suddenly burst out laughing: a good, loud laugh at the next song chosen by Anton. Like most Russians his age, he grew watching the Sanremo Music Festival on Soviet TV and he loves the music of Adriano Celentano. Nothing weird about that, but the coincidence of his unlikely choice is irresistibly funny. Here we are, hermetically sealed at the top of a rocket, the launchpad evacuated. After years of training, months of preparation, weeks of exams and formal events and the frenzied days before the launch, we are finally enjoying a moment to ourselves, out of everyone else's reach – and my commander, who understands not a word of Italian, has chosen a song that begins: 'No use ringing / No one in here will open up / We've shut out the world and its chaos.'

It's not just Celentano: there's amusement in the air. We joke as we've always done, from the moment we became a crew two years ago. We laugh at our uncomfortable, scrunched position, which must be most painful for Terry – like me, he is lying in a poky side seat, but he is much taller. We make fun of his attempts to distract himself with sudoku – he can't solve any of them. We joke about our joking, and the fact that we're only fifteen minutes away from our launch, and there's no sign in here of the solemn atmosphere that everyone out there must be imagining. Maybe we're a particularly cheeky crew. I don't know. Maybe other crews have exhibited more gravitas at this moment. Maybe . . . but I'm not convinced. Taking yourself too seriously is considered a major sin for astronauts, and it's something of a faux pas to forgo an opportunity to downplay a situation.

At ten minutes to launch, we receive instructions to close our helmets, and this simple action refocuses our attention. I lower the visor and use the mirror on the forearm of my suit to make sure nothing is caught in the seal. Once again, I check all systems and fill out a table I've prepared to record the status of the main parameters before lift-off. Then, in the procedure checklist, I turn to the launch and ascent page.

It's quiet on board now, but before long the neutral voice of 16-3rd breaks in to announce: 'We are GO for launch. Everything is proceeding to plan. Transfer to on-board control. I'll transmit the progress of the launch via radio.' Anton confirms that we're ready. Unless there's a technical failure, this rocket will launch in a few minutes, whether we're ready or not, but I'm pleased to hear him say it more decisively than usual: I can sense his mind

233

and spirit are present and alert. Or maybe I'm the one who feels like that – present and alert – every fibre of my being straining to feel the rocket's lift-off, all thought focused on the distilled essence of life consumed in this glimpse of space-time, my spirit quiet and calmly open to what will take place in the next nine minutes and the next six months.

There were just a few seconds left before the launch, and each radio announcement comes fast on the heels of the last one: *'Zemlya-bort,'* says 16-3rd, telling us that the last arm of the launchpad has been retracted – the last umbilical cord cut. *'Pusk,'* start. Kerosene and liquid oxygen begin flowing into the combustion chambers. *'Zazhiganiye,'* ignition. The propellants begin to burn. From now on there's a rapid increase in noise and vibration as the engines get up to full thrust. From the bowels of the rocket comes a sudden jolt, and at midnight, one minute and fourteen seconds – exactly as planned – 16-3rd confirms lift-off over the radio. *'Poekhali!'* Let's go! Anton cries out, just as Yuri Gagarin did in his time, and we join him in an exultant cry. I don't know what to do with my happiness. It's so big it's getting in the way. I feel it spilling out in a smile I couldn't stop even if I wanted to.

On this perfect night, the Astrey crew is leaving the planet.

28.

The goal of life is to make your heartbeat match
the beat of the universe.
     Joseph Campbell, *Reflections on the Art of Living*

Soyuz TMA-15M, 24 November 2014

The rocket plucks us from Earth in a firm but gentle grasp. Every
now and again it adjusts its attitude in order to keep us on the
planned trajectory, and then we feel a dull thump accompanied
by some oscillation, but all told, the vibrations we feel are light.
The liquid-propellant engines on the Soyuz don't have the fury
of the solid-fuel boosters on the Space Shuttle, which shook the
screens in front of the astronauts. It must seem to Terry, accustomed to that violent lift-off, that we're not moving yet.

But we're moving, and how! With every second that passes
we become lighter by one and a half tonnes, as the propellants
burn and are expelled through the nozzles. As a result, the acceleration progressively increases, pressing us into our seats with
increasing force. It's not unpleasant, at least for now – quite the
contrary. It feels like my body is sinking into the rocket, becoming one with it.

'Thirty seconds, and engines are stable,' announces 16-3rd via
radio.

Anton responds mechanically, 'Everything's in order on board.
The crew feels fine.'

We must be about to exceed the speed of sound, if we haven't
done so already. It's a significant transition in terms of the rocket's structural integrity, but we don't feel a thing inside. We're
leaving the sonic boom behind.

In the checklist the launch procedure is condensed onto a single page, a simple graph depicting the ascent and the main events as a function of time elapsed from lift-off. After 114 seconds, for example, at an altitude of 40 kilometres, the launch escape system will separate. We have no indication of our speed or acceleration on the control panel, nor any telemetry data on the rocket's firing. The physical sensations, the noise, the growing force that presses us into our seats – all this is reassuring, but the only objective information on our ascent reaches us via radio.

'Sixty seconds. The attitude is nominal,' says 16-3rd.

I know that in the next minute the G-force will rapidly increase under the action of the five engines thrusting us upwards – they are now pushing a rocket which is already roughly 100 tonnes lighter. As a precaution, I begin using the breathing technique we learned in the centrifuge, stiffening my rib cage. The vibrations are increasing in intensity, and we start bouncing around as if we were riding a bike at full speed over an uneven road without shock absorbers. Olaf, hanging from his cord, is swinging around crazily. My procedure checklist is shaking. I'm holding it only halfway up, and the top of the checklist is bending forward under the effect of the acceleration. To flatten it so I can keep reading, I have to stretch out my arm, and it's now difficult to lift it and control its movement. As we approach first stage cut-off, we're pinned to our seats by a force equal to more than four times our weight.

This is it, I'm thinking, when I feel the first blow. But the rocket continues to push with the same force. It was just the separation of the launch escape system. The actual first-stage cut-off is the next jolt, a few seconds later, the one that accompanies the shutdown of the four lateral boosters. I feel myself shoved forward, but trapped by the straps and held firmly against the seat. Our velocity continues to increase under the thrust of the core engine; however, the acceleration has diminished suddenly and with it my perceived weight. It's momentarily disorientating, but the brain adapts quickly. From 16-3rd, we get confirmation that the second stage is working perfectly.

The ascent chart in the checklist shows the next major event: the separation of the fairing, the aerodynamic shell that has protected us as we've gone through the atmosphere's dense lower layers. I grab the torch stuffed between my kneeboard and my leg and point it at the small, round window on my left. The metal shell is out there, a few centimetres away. I wait. The rocket is shaken by two strong jolts that signal the ignition of the pyrotechnic charges, and with a sudden leap the shell disappears into the darkness. I can't say I saw it leaving: one moment it was there and the next it was gone. It's still night outside. I go back to focusing on the launch.

As forecast, the 'Guided Descent' button lights up on the panel, which means that from this moment on, if there should be a launcher failure, we could still hope to return to Earth in a controlled manner rather than ballistically. After the automatic separation from the rocket, we would have three minutes to decide – three minutes to look outside and confirm that our roll angle is less than 45°. In the absence of that condition, any attempt to make a controlled re-entry would result in our literally being killed by the G-force. We've spoken about it a lot over the past few days. Given that we are not trained to evaluate that angle and that it's night outside anyway, we'd definitely opt for a ballistic return, with its unpleasant, even dangerous dose of Gs, but at least not inevitably fatal.

We're now past four minutes of flight time, and the second-stage engine will soon shut down. In this final phase, the smooth engine firing is interrupted by moments of irregular, jerky vibrations. They're not terribly comforting, to be honest, but 16-3rd keeps reassuring us via radio that everything is nominal. At the appointed time, the engine shuts down, thrusting us forward against our straps. With a bit of juddering and thudding – and our excited cries – another phase of our ascent to orbit is concluded.

I feel the push from the third stage and the voice of 16-3rd, which lets us know that the engine has ignited as it should. We're at an altitude of about 160 kilometres, according to the graph in the procedure checklist. Whatever happens from here on out,

I am officially an astronaut. In the first phase of our ascent, we climbed along a roughly vertical trajectory to get through the dense atmospheric layers as quickly as possible. In the remaining four minutes, the thrust of the third stage will serve most of all to help us gain the horizontal velocity we need to stay in orbit. Getting to space is easy, as they say; the difficult thing is to acquire the speed to stay there.

Should there be a failure in this final phase, our landing would be a treacherous one. The last line of the checklist page about the ascent sums it up in two words: mountains, and then the sea, it says. The immense and relatively welcoming Kazakhstan highlands are now behind us. 16-3rd continues to transmit radio updates on engine firing, Anton in turn assures him that everything is fine on board, and I periodically check the parameters of our spaceship. Up to this point, it's been a textbook launch. My visor didn't even get fogged up.

Thirty seconds before the scheduled engine cut-off, I raise the protector over the KO command, what we call the 'separation contact'. When we're in weightlessness, a number of lights on the control panel should light up to confirm the separation of the launcher, and if they don't, I'll have to send the command manually within ten seconds. It's a simple action; all the more reason why I don't want to make any mistakes. If we don't get to the Space Station tonight, it certainly won't be because I didn't send the KO command in time.

However, there's no need. The Soyuz does everything itself. The engine of the third stage shuts down, the invisible hand pressing me into my seat releases its grip, and I'm thrust against the straps. I watch as Olaf begins to twirl around, and a cloud of white particles goes by my window: debris driven there by the engine's cut-off. We detach from the third stage with a shudder, and a series of dull thuds alert us to the extension of the antennae and the deployment of the solar panels. A few seconds . . . and our Soyuz is ready to fly on its own.

A sudden silence falls over us. Of course, the ventilation system keeps humming like a swarm of bees behind my head, but after the racket of the engines and the succession of explosions from the charges, it feels like silence has descended on us. In their thick gloves, my hands are dangling at about eye level, as if they weren't attached to me. In an immediate flip that flies in the face of millions of years of body memory, I have to make an effort to hold them against my body. I stare at them for several moments, entranced, before Anton's voice reminds me that I have other duties. 'Let's go to page 33,' he says, and it's a pleasure to follow his order, skipping over four pages of procedure, the ones we would have used if there'd been a launcher failure. Our mission has barely begun, but we've already got past one of the two riskiest phases: the launch. The other will be our re-entry, now six months away, when we'll have to release in a controlled way all the energy the rocket stored in our little Soyuz in the form of altitude and, above all, velocity.

Now that we're in orbit, we begin our pursuit of the Space Station. It went over us a couple of minutes before the launch, when we were still in Baikonur. We've entered its orbital plane, and now it is about 3,000 kilometres ahead of us to the northeast, above Mongolia, with a phase angle of 25°. In the next six hours we'll bring that angle down to zero. Relative to the ISS, we're on a lower orbit; therefore we fly faster.

'Astrey, TsUP from Moscow. How do you hear me?' Now that the launch is over, responsibility for the mission has passed from Baikonur to TsUP. Dima went back to Moscow a couple of days ago to be ready on site. It will be his familiar voice we hear for the next few hours. He pronounces his words calmly and walks us through the first operations. Time is short now, as the Soyuz will soon start preparing automatically for the first two engine burns. It's Anton's duty to supervise these operations. I have to check that the systems work properly and carry out the leak check. Luckily, I feel well. I'm moving my head cautiously, of course, unwilling to tempt fate, but I'm not feeling the least bit nauseous. I retrieve my pencil from mid-air, where it's dangling from a cord, and I write down system parameters and pressure values for the leak check in the appropriate chart.

'Samantha, look outside!' Terry urges me a short time later. I was so absorbed in my work that I hadn't noticed it: the orbital night is coming to an end. The last stars are fading, and the world outside is tinted blue and red. It's my first dawn in space. Just a fleeting glimpse in my uncomfortable position, as I negotiate with my confused vestibular system and continue the checks. But what a sight! I look at it again a few minutes later. As we fly towards the Sun, the mantle of darkness recedes below us. There's the Earth, down there: blue, as it should be, with sprinkles of white that cast long shadows over the sea.

We're about to lose Russian radio coverage. Dima makes a last, rather formal radio call before bidding us goodbye. 'Astrey, the spaceship is in orbit as an artificial satellite of the Earth. Antennae and solar panels are deployed. The orbital period is 88 minutes, the inclination is 51°, the altitude is between 199 and 242 kilometres. You have permission to loosen your straps and raise your visors. Until the next radio contact–'

'Ciao, Dima,' I think. 'We'll just circle the Earth and then we'll speak to you again. Don't go very far!'

I turn off the radio and to make sure I remember to turn it back on, I set the alarm for one minute before the next pass over the Russian stations. Half an hour later, I can definitively state that our Soyuz does not have any leaks, and I announce that in accordance with procedure, we can take off our gloves. Terry and Anton don't need to hear it twice. The gloves are a serious hindrance. I go back to reviewing my own sensations: everything's OK overall, but I'm beginning to feel an irksome pressure in my head, no doubt due to the fluid shift, or the redistribution of body fluids that happens during weightlessness. My vestibular system is also starting to play tricks with me. I have a clear sensation of being in forward freefall, a feeling that I would crash onto the control panel if my straps were not holding me in. I decide to tighten them, allowing myself less space for floating, and that seems to help.

When I finish checking the systems, I join Anton in monitoring the engine burn. The first one is now imminent, and it will last a minute and a half, taking us that bit closer to the Space

Station. It's a slight thrust, but it's enough to upset my vestibular system, which is already being tested: I force myself to look straight ahead, without turning my head, and the sense of disorientation slowly disappears.

After all the excitement, there's not much to do now until the next engine burn in about an hour. The Sun goes down, marking the end of my first orbital day, and the first stars come out. Terry suggests that we turn off the light for a couple of minutes in order to get rid of the only light pollution there is up here. I spy the Milky Way, and I recognize Cassiopeia. They're as far away as they are from the Earth, but up here it seems like we could have a conversation. Ah, what a silly thought! People have always spoken to the starry sky. And yet there's something poignant about contemplating the heavens from inside this metal box, a tangle of technology – not even that advanced – that allows us to be up here, in space, where humans are not meant to be.

I go back to focusing on the systems and I notice something suspicious about the pressure in the descent module. It goes down, comes up, then goes back down. If we were in the simulator, I'd prepare for a depressurization, but the actual sensors aren't so precise or clear, so I shouldn't rush to conclusions. I do alert Anton, and we decide to continue monitoring the situation . . . It's so amusing to send the procedure checklist spinning in the air! It's really something, this weightlessness; it's just that this pressure in my head is a nuisance . . . What a spectacle on the Earth below: there's a storm, a furious succession of flashes like strobe lights in a disco . . . Meanwhile, the pressure values keep going up and down, but the trend is not downward, and in any case, if there were a real problem we'd feel it in our ears . . . It's time for the second engine burn. As always, Anton is watching our attitude and acceleration, and I'm keeping an eye on the operating pressures. Everything is going just as it should . . . Terry starts using a hand-pump to transfer the condensate from the air-conditioning system into one of the storage tanks . . . The alarm rings, and I turn the radio back on.

'Astrey, TsUP from Moscow. How do you hear me?'

Later, after two orbits and two engine burns, reacquiring the radio signal is more difficult. This time we expect to establish contact with the Space Station. It's now barely 200 kilometres away, and it should act as a radio link with Moscow via Houston, so we won't have to rely on the infrequent Russian stations coverage. We're now an hour and a half away from docking. At last we hear Dima's voice, weak and unintelligible. Conversation is impossible, but Anton keeps on transmitting, periodically reporting our distance and relative velocity. We're on the braking parabola: in a few short series of engine burns, the computer will take us to the same altitude as the ISS, progressively decreasing the speed of approach. At around 100 kilometres, the radio contact stabilizes. The first voice we hear from the Space Station is Sasha's. 'C'mon guys, hurry up! We're waiting for you.'

We begin to feel how tired we are, and we try to help each other to stay awake. During one of the long pauses between one engine burn and the next, Dima thoughtfully enquires whether we've all had the chance to go into the orbital module. This is his roundabout way of asking if we've been able to use the toilet. There was an interval marked out for this a couple of hours after the launch, but Anton took a long time over the installation of the video camera GoPro, which will film the docking process through the front window. There wasn't any other time for it: in the manoeuvring phases we can't move from our seats because the computer would interpret any shaking as a failure of the attitude control and would categorically refuse to start the engine. But during the previous radio contact, Dima indicated another possible interval before the start of the final ISS approach phase, so Anton can now confirm that flight engineer no. 2 has indeed been in the orbital module. 'Successfully,' adds Terry, his voice sleepy but clearly amused.

'Successful, right, Terry?'

'Yes, one hundred per cent successful!' Terry gloats.

'Empty,' Anton adds, and we titter, along with Dima and probably various other control centres all over the world as soon as the conversation is translated into English, which is what happens in real time with all Space-to-Ground communications that go through the Space Station.

For the second time tonight, I envy Terry and Anton their empty bladders. There wasn't any opportunity for me to go into the orbital module, but I didn't really expect that much. There's almost never time for the flight engineer to go. All I want is to be able to pee in my nappy. I don't ask for much, right? And yet I can't. No one was able to explain this inconvenient and involuntary reluctance, but it's a well-known problem, so much so that various astronauts advised me to try peeing while lying down in the bath, in the hope that this would act as positive conditioning or at least help to lose any mysterious inhibitions. I did it a few times – successfully, as Terry would say. But right now, no matter how hard I try, something unconnected to my conscious will is stopping my bladder from emptying. I'm trying not to worry, and I tell myself that the problem will resolve itself later, when we can use the Soyuz toilet, but I know very well that this story could have another epilogue, featuring the catheter. It's already been used, I'm told, and more than once. During our medical training, Terry and I got some practice inserting the catheter in someone else, using a dummy, and I even gracefully accepted the offer of additional training on how to catheterize myself, taking a kit home with me in order to practise. But things are very different up here: everything floats, and, as a rookie in space, I would find it difficult just to stay still.

Anton distracts me from my physiological worries to show me the image from the external video camera. There it is, the Space Station. In this absolute blackness, empty and hostile, our Soyuz is pointing securely at our only conceivable destination, this outpost of humanity where Sasha, Yelena and Butch are ready to welcome us with a hot meal. By now we're in constant contact, and our transmitter is kept on so that the ISS and other control centres on Earth can listen in on everything we say. On our side, we are ready. The docking probe is extended, and the Kurs system antennae are working perfectly. At 3 kilometres away, there's yet another automatic engine burn, and from now on we could also take manual control if necessary. I shake off my sleepiness, forcing myself to make a final effort to concentrate: if there were a failure at this stage, we'd have to act very quickly to effect the docking.

We're perfectly aligned with the port of the MIM-1 module, slowly but surely approaching the ISS, when I allow myself a quick look outside. The moment isn't chosen randomly: I'm looking for the Space Station. Not the indirect, black and white image from the cameras. I want to see it with my own eyes, in colour. I'll never be able to see the whole thing, but I'm waiting for part of it to appear in my window. Some of the veteran astronauts have told me to look for it in the last 40 metres. I've loosened the straps as much as I can and I'm floating over my seat. When I turn towards my window, the first thing I see is one of the small solar panels on our Soyuz. Nothing new there: I've seen it before, but something draws my attention to the edge of my peripheral vision. I turn my head slowly and feel a jolt of astonishment and joy. Maybe my pent-up emotions are bubbling over. Maybe my tiredness is finally getting the better of me. Maybe it's simply that I wasn't prepared for such sudden beauty. 'Oh my god!' I exclaim, and Anton immediately orders me to be quiet. He's right, of course. The radio transmitter is on, and on Earth they cannot know that everything is fine, that there isn't some problem – other than an overabundance of happiness. The Space Station is right there, outside, mighty and resplendent, lit up with orange light, warm and alive, in which its huge solar panels seem to burn, incandescent. When I turn to look outside a few seconds later, the orange blaze has disappeared. The Space Station is now bathed in a cold, metallic light – and then it's swallowed up by the darkness. Another orbital night has begun.

I realize that I've just had my first glimpse of my new house, precisely in those few fleeting moments of orange glow, in the transition from day to night. And I have the absurd sense that the universe – so indifferent to the fortunes of humankind, let alone those of a single person – that universe wanted to give me a gift tonight.

29.

> A person's life consists of a collection of events, the last
> of which could change the meaning of the whole, not be-
> cause it counts more than the previous ones but because
> once they are included in a life, events are arranged in an
> order that is not chronological but, rather, corresponds to
> an inner architecture.
>
> Italo Calvino, *Mr Palomar*

International Space Station, 24 November 2014

'*Kasaniye*, contact!' exclaims Anton, with an enthusiasm that betrays his emotion and relief at having arrived at our destination
without any hitches. He quickly reassumes his neutral, professional tone to relay his commentary, confirming, via radio, that
the proper lights go on and off in the proper sequence, indicating that the automatic docking system is functioning correctly.
We're actually transmitting the image on the screen to Earth, so
the TsUP controllers are seeing exactly what we're seeing, but
they are careful not to release Anton from his duty of reporting
everything by radio. This commentary is sacred for the Russians,
and may be a relic from a time when video transmission was less
reliable.

Right after confirmation of contact, or confirmation that our
docking probe has touched the reception cone of the MIM-1
module, the attitude thrusters go on for a final, brief push, until
the capture signal is illuminated: the probe's head has reached the
bottom of the cone and has been captured by the locking latch-
es, the so-called 'first mechanical coupling'. Our docking probe is
now being retracted, a process that will take ten minutes. It's as if

we were trying to draw the 400-tonne ISS towards us. More realistically, we are the ones being drawn closer and closer, until the hooks on the docking system close, sealing the interface. Now we are one with the Space Station.

That is, mechanically we are. In practice, we're still separated from the Space Station by two hours of meticulous leak checks. We have to confirm that there are no leaks in the tunnel that now joins us to the ISS and that will allow us to float over to Sasha, Yelena and Butch, once we open the two hatches.

'Holy cow, there's a Space Station out there!' Terry says all of a sudden. He already has that nasal tone, so characteristic of space, which is caused by the fluid shift.

'Oh my god, the sunrise! Holy, holy cooooooow!' I exclaim in turn. I'm ecstatic and I look outside, once more abandoning professional reserve for aesthetic rapture. Now that there's no rush I have the time to abandon myself to the silent spectacle of light and metal beyond our window: the Space Station is emerging from orbital night. Splashes of light from the low Sun's rays hint at the complex geometry of the space outpost, which is still mostly shrouded in darkness. As the Sun climbs higher, its light outlines the contours of the various modules in their entirety. 'Columbus, the Airlock, PMM,' Terry says. On my side I catch sight of the Russian segment, and also the large radiators. I'm fascinated by every detail: after all these years studying it, the Space Station is right there, only a few metres from my window. I've so wanted to come to this place that I can't help but love it, all the more so knowing what a privilege it is to be here. Only 215 people have been to the Space Station. I'm about to become the 216th.

Before that glorious moment, however, I have to take care of something much more prosaic, which is becoming more and more of a problem. Anton has already gone into the orbital module to start the long process of leak checks.

'Er, Anton . . . could I take care of that? I have to use the toilet.' I keep the mic away from my mouth. It's a somewhat unusual request, since leak checks are usually the responsibility of the commander, and in training it was always Anton who simulated this part of the procedure. But he doesn't bat an eyelid, requesting TsUP's authorization straightaway.

'Can Samantha go into the orbital module to start getting changed?'

No problem, as long as I keep transmitting pressure values. TsUP relies entirely on us for these readings, since we take them on a mechanical pressure gauge, a manual instrument that doesn't send any telemetry. We even have to tap it a few times before we start to measure, in case the arrow gets stuck. So, for the first time, I leave my seat and start floating, pushing off awkwardly towards Anton. Weightlessness is fantastic! It's all been worth it to this point, if only for this sensation, to slip through the open hatch and climb into the orbital module without the least bit of effort. What a terrestrial verb: 'climb into'! A few days from now – weeks, maybe months – I may stop thinking in these terms. For now, I am acutely aware that Anton and I are floating above the circular opening on the floor of the orbital module: you had to be careful on Earth, not to fall through it into the descent module. Every now and then, Anton grabs me and helps me to stabilize my position. He's already at ease up here and knows how to move. Floating must be like riding a bike: it stays with you as muscle memory.

Before he leaves me on my own in the orbital module, Anton helps me to take off my Sokol suit. We always help one another so we don't waste time with hooks, laces or zips, but this time I'm mostly pleased that Anton is helping me to stay still. After a quick handover concerning the leak checks, he nimbly pushes himself towards the descent module, landing elegantly in his seat. He pulls the hatch to, in order to give me the privacy demanded by the situation. I take my bearings for a moment. Much as I long for the relief of the toilet, the most important thing is to perform the checks. Right now we're testing the pressure gauge itself for leaks, and I need to transmit an update on the pressure values every minute. After that, I'll need to operate a series of valves to check the different air volumes in the tunnel that joins us to the Space Station. As astronauts we're ready to joke about anything and everything, but we always take valves seriously. When we hear the hissing sound of a pressure equalization, we like to be sure – absolutely sure – that we're not hearing the air we're breathing as it escapes into open space. I haven't worked with these valves since my initial lessons as flight engineer, but happily, the system is astronaut-proof: I'm sure I remember that it's impossible to simultaneously open a combination of valves that will allow air to escape outside. Nevertheless, I quickly study the diagram that shows the pneumatic connections and is posted beside the valve block, to reassure myself that I understand the procedures well. In the meantime, Dima calls over the radio for a pressure reading.

'TsUP Moscow, Astrey 2, at 6.56 the pressure gauge is indicating stable pressure, 3 millimetres of mercury.' One minute before the next reading. Clumsily, I turn around in mid-air. With a push here, a kick there, I take off my leggings and my nappy and then, hanging half-naked over the partially closed hatch to the descent module, I transmit the pressure reading over the radio. I'm well aware that my situation could be perceived as slightly comical and a bit embarrassing, but on the other hand, I'm on my own. When it comes down to it, no one can see me; they can only hear me. All I have to do to preserve my dignity is to keep using the radio as if I were completely at ease, with nothing in my life to worry about besides reading a pressure value every minute. I've learned over the years that the best astronauts, just like the best pilots, know how and when to play the actor. So I transmit the pressure gauge readings on the dot every minute with the tone of someone who's been checking pressures in space for years, and miraculously, between one measurement and the next, I manage to use the rudimentary toilet on the Soyuz, marvelling that the airflow actually manages to take everything in the right direction. And with this demonstration of space hydraulics at work, the spectre of having to be catheterized recedes. The rest of the mission will be a breeze.

A couple of hours after docking, we've finally established that there are no leaks, and Sasha confirms by radio that the hatch of

MIM-1 is now open. Our own hatch, the final barrier between us and the Space Station, is still firmly in the closed position, locked not only mechanically but also by the overpressure in our Soyuz – barely 30 millimetres of mercury, the difference in atmospheric pressure between a sunny day and a cloudy one, but enough to keep us from opening the hatch at all. We therefore open the equalization valve, which allows the air to flow through to the Space Station, and we wait until the pressure is the same on both sides.

The arrow on the pressure gauge moves, but not as fast as we'd like. We all keep looking at the clock. An important deadline is approaching: the end of coverage on the Ku frequency band. We'll still be connected on the S-band, the one that allows for audio communication and the transmission of critical telemetry, but video is sent on the Ku-band, which is subject to longer and more frequent interruptions. Well, we'll be losing the coverage in a few minutes. And since it's unthinkable that we should make an entry into the Space Station without sending the images to Earth, where our families and friends are anxious after a night of waiting, we have to hurry in order to open the hatch in time – or else resign ourselves to another twenty minutes of waiting.

'745 millimetres of mercury,' Anton reports, and he adds an implicit request: 'Almost the same as on the Space Station.' We still need a couple more millimetres, to be honest, but TsUP gives us authorization to proceed with opening the hatch. At this point, things start to speed up. I assist Anton with the installation of the crank handle. A couple of turns, and the hooks release their hold. Anton tries to open the hatch, but the tiny bit of remaining overpressure keeps it shut. The knocking coming from the Station tells us that Sasha is trying to help us from his side. He pushes, Anton pulls, Terry and I wait, by this point resigned to a delay of another twenty minutes until we recover Ku-band. But suddenly the hatch begins to open.

'Ready, Samantha?'

I'm somewhat nonplussed. We never talked about who would go in first, but I certainly didn't expect it to be me. Anton is the commander of the Soyuz and Terry will be the commander of the Space Station. I ought to thank them and politely decline, but in the rush of the moment it feels like arguing would not be appropriate. And I must confess, this makes me happy. As soon as the hatch is open wide enough, Anton gently pushes me through to the Space Station.

'Go, Samantha, go!'

It's like a second birth. I squeeze from the cramped quarters of our little spaceship into my new life, floating towards the human beings who inhabit it. It's an intense moment, one of those rare points of connection between past and future that propel you toward what is yet to come while shaping and completing everything that occurred before in an instantaneous backward action that flows upstream through time.

I hug Sasha, Butch and Yelena tightly, at the same time trying to move along the narrow MIM-1 module so I won't block Anton and Terry behind me, or the cameras filming the event. We need to get into the Service Module right away for the Space-to-Ground chat with our families. Yelena leads the way along the FGB, another long, narrow module that has lots of handles for pushing yourself along. We emerge into the Service Module, and for the first time I'm able to float in a large space. I can't find any handholds: on all the walls that I can reach there are cameras, lenses, pens and other objects attached with Velcro. Then I see a window, small and round. Somehow I manage to grab hold of a handrail and I find myself with my face against the glass, without even knowing how. A patch of Earth glides by beneath me, as lazy and majestic as a river on the plains. I watch red, sandy expanses alternate with and wrap around dry, black reliefs, a harsh landscape immediately softened by a host of small clouds. I can't make out any cities or streets. Those who wish to look carefully at the earth should stay at the necessary distance,' says the Baron in the Trees.* I have done my best.

Sasha reminds me that Baikonur is waiting for us to make a

---

* Italo Calvino, *The Baron in the Trees.*

video call and suggests that I come to the back of the Service Module, where Anton is doing somersaults, as gleeful as a child on a merry-go-round. Yelena helps me to get in position behind the table while Butch moves elegantly over my head and then comes feet down behind me. Soon we're all lined up, the new arrivals wobbly in the front, the space veterans floating steadily behind us. Somehow I find myself wearing headphones and a microphone. It's all been so quick! It's only been about five minutes since we got here.

All at once they call us over the radio: 'Station, Houston. Are you ready for the event?' Anton starts to answer in his halting English, but Terry jumps in, surely delighted to hear his own language through the earphones after all these weeks.

'Station is ready for the event,' he confirms.

'Baikonur, this is Houston. Please call Station for a voice check.'

So begins every public in-flight call with the Space Station. This time, however, there aren't any students, journalists or VIPs on the other side – just our families, tired and emotional. It's naturally a bit embarrassing: what do you say to each other during an event streamed worldwide? The adorable Kira, too little to worry about it, gets us all smiling.

'Papa, I saw the rocket this time. This time I didn't fall asleep!'

'Well done, Kirichka! I'm looking forward to seeing your photos of the launch.'

'*Poka* Kira, goodbye,' we all say at once.

'I'll call you again soon,' Anton adds.

That's right: there's a phone on the Space Station, a VOIP system that allows us to hook up to the Earth's phone network, and the cherry on top is that it works in only one direction: we can call anyone, but no one can call us.

While Terry is talking to his daughter, we lose Ku-band for about a quarter of an hour. Butch, Sasha and Yelena have prepared a little feast for us. There's a small oven compartment hidden in the table, and they're heating up assorted cans of meat and fish in it. Butch brought some NASA packets from Node 1,

and he pulls these out of a second oven, which is a sort of metallic suitcase with an electrical heater. In a minute, the table is set
with packets and cans held fast by duct tape, folded sticky side up.
There's a rubbish bag tied to a corner of the table, tightly closed
with rubber bands. Someone offers me a packet of water and I
immediately prove to be a rookie by leaving the straw open after
drinking. We're all euphoric, despite our tiredness and the uncivil
hour. It's about six in the morning on the Space Station, and it's
been a very long Sunday, even for Sasha, Butch and Yelena. It's
not over yet: we still have a few hours' activity ahead of us – the
emergency briefing, for example – before we can rest. Anton's
crew quarter is in here, right beside the table, opposite Yelena's.
Terry and I will live with Sasha and Butch in Node 2, where there
are four crew quarters arranged like spokes.

When communications are re-established, it's my turn to talk
to my family. We chat, juggling emotions with embarrassment,
before Lionel casually defuses the situation. 'I have a question
for you, Terry and Anton. Now that you're up there, can you tell
me the answer to the ultimate question of Life, the Universe and
Everything?'

It's a relief to end with a joke. 'The answer is forty-two!'

A couple of hours later, when I'm ready to collapse with tiredness, Terry calls me urgently. 'Samantha, come quickly!'

I can guess what it's about. Terry has jokingly forbidden me
to look out of the Cupola without his permission; he wants to
choose the perfect moment.

I feel like a child who's about to unwrap a long-awaited Christmas gift: excited, but also anxious, in case it's not as wonderful
as I imagined it to be. I follow Terry through Node 3 and then
plunge through a wide opening in the floor, to find myself in the
Cupola at last, the Space Station's fabulous little veranda. I'm
surrounded by windows that end level with my chest. The main
robotic workstation is beneath them. The rest of the lower wall
area is taken up by countless cameras and lenses. I turn around
to look at the panoramic view through the six peripheral windows, and then look up to the large circular window pointed

towards the planet's surface. The Earth is a black hole. Literally. We're surrounded by stars, but below us is complete darkness. Terry has chosen the moment well. We're flying over the Pacific Ocean, an immense expanse of water, with not a single light to suggest the presence of humanity. Eastwards, the Earth's contour is barely outlined by a faint blue arc, like a sliver of sky bursting into a dark room through a thin, arched crack. Little by little, the crack opens, and the blue crescent stretches across the horizon like a caress. A few minutes later, the imminent dawn announces itself with a slender stroke of orange that gradually expands in an ever wider, bright embrace. The Sun arrives on the scene, shabbily dressed at first as a blotch of colour, a compact egg yolk on the horizon, before its warm light overflows and spills onto the black canvas. It's a feast of light: the horizon is layered in infinite shades of blue and orange, swept away at the centre by the shining yellow Sun. Then, as if the split in the black canvas were finally ripped apart, I am struck by the blinding force of a fully risen Sun, which unfurls its light over the surface of the planet, beating back the darkness.

At last I go to bed, reluctant to let go of this long and perfect night, but with the warmth of the newborn Sun on my face.

30.

Our travellers had been fully prepared for such a phenomenon, yet it struck them with as much surprise as if they
had never uttered a scientific reason to account for it. They
saw that, no longer subject to the ordinary laws of nature,
they were now entering the realms of the marvellous. They
felt that their bodies were absolutely without weight. Their
arms, full extended, no longer sought their sides. Their
heads oscillated unsteadily on their shoulders. Their feet
no longer rested on the floor.

Jules Verne, *From the Earth to the Moon*

International Space Station, 24 November 2014

I try to turn the light on, but I can't find any reference points to
guide my hand in the dark. How difficult can it be to find the
switch, when my crew quarters are about the size of a phone box
and I know very well where to look for it? Feeling around blindly, I end up bumping into the door with my shoulder, which un-
locks and opens part way, letting in a bit of light. I get it now: I'm
topsy turvy. It seems incredible given these cramped quarters,
but I somehow managed to flip over in my sleep. I want to say
'upside down', but it doesn't make sense. 'Down' with respect to
what? My crew quarters are in the floor, or what we consider the
floor: the area where you put your feet, if you want most of the
signs in the Space Station to be correctly orientated for your eyes.

Butch sees my head poking through the floor, and he comes
towards me from Node 1, floating rapidly through the Lab with
a couple of well-placed pushes. 'Good afternoon! How did you
sleep?'

Very well, I'd say, and no, I didn't tie myself to the wall with bungees. It's a bit early to know for sure, but I think I'm going to like sleeping like this, floating. Butch shows me how he sleeps when he wants to feel some pressure on his body: with his back against one wall and his legs pushing against the opposite wall. I take note.

I've woken up hungry, so I ask Butch for a quick handover session on food preparation. I follow him clumsily into Node 1, fumbling like a child just learning to walk. He assures me that in a couple of weeks I'll be twirling around like a ballerina. I just hope I don't break anything in the meantime. In Node 1, the food packets are organized by type: breakfast, meat and fish, sides, vegetables and soups, fruit, including dried fruit and nuts and finally snacks and sweets. It's mid-afternoon, but I feel like having breakfast. Butch points out a makeshift cupboard, a big white box with a Velcro closing, labelled 'Breakfast', in fact. There are a dozen of these boxes and a similar number of smaller metallic containers on the opposite wall, and together they form a sort of improvised pantry, clearly the brainchild of some of the earlier crews rather than an actual design. I take a packet of porridge and another of scrambled eggs, both dehydrated, and dip into a box on the opposite wall containing hot drinks, including tea and various types of coffee: sweetened or unsweetened, with or without milk. While I'm looking for some black coffee, a handful of packets start flying away – and while I'm trying to grab those, I let go of the scrambled eggs; and while I'm trying to catch the eggs, I end up turning over in mid-air and losing my sense of direction. Up, down, forward, aft . . . It's all so confusing.

Holding the wayward coffee packets, Butch watches me with amusement and asks me how I feel. Nauseous? Headache? Nothing, luckily, but I've got a sickbag in my pocket just in case. The pressure in my head is still there. Butch had the same sensation at the beginning, but he tells me it went away after a week. I follow him to the water dispenser, which is on the ceiling in the middle of the Lab, and he shows me how to maintain my position by putting my feet in a restraint conveniently fixed to the wall. It's a little like putting on flip-flops attached to a handrail.

The food packets have hooped pieces of Velcro stuck to them, just like every other portable object on the Space Station. Meanwhile, the many successive crews on board over the past fifteen years have left their trace in the form of pieces of fuzzy Velcro on free surfaces all over, but especially in a few strategic or much-used places. There's one right next to the water dispenser. I put the porridge there while I rehydrate the eggs according to the instructions on the packet: 150 millilitres of warm water. It's hardly more than lukewarm so you won't burn yourself, while the cold water I pour into the packet of porridge is not really cold, just room temperature. Happily, there's a small fridge in this rack. When it's not being used for experiments we have permission to put drinks packets inside. Normally I only drink water, but on board there are also fruit juices and shakes, all powdered. The only things we don't have are carbonated drinks.

While my food is resting for fifteen minutes, according to instructions, I venture into Node 3 to use the loo. Compared to the one on the Soyuz, which is composed of only a funnel and a receptacle attached to a flexible tube, the toilet in Node 3 is more like a normal Earth toilet, as long as you don't pay any attention to the two panels with warning lights and switches that betray the presence of a jumble of pipes, pumps, fans, tanks and filters hidden inside the rack. A Japanese toilet is boring in comparison. To ensure privacy, there's a cabin around everything, which cuts the space of Node 3 in half. It's got an accordion door and on the walls are packets of wet and dry wipes, a pack of latex gloves and a bin bag for used hygienic products. Instead of a toilet bowl there's a chunky metal cylinder for solid waste, and on top of that there's a small white seat which is kept covered when not in use. Butch has already advised us on the most efficient and stress-free technique: though there are various footholds to help you stay in position, it seems that pushing against the ceiling with your hand is the easiest method. In any case, I'll try some other time. For now, I just take the flexible hose off the wall, which has a conical yellow funnel on one end for pee. I know that somewhere on the Space Station there are differently shaped funnels intended specifically for women, with forms better adapted to the female anatomy. They also have holes on the sides, so they can be pressed against the body without blocking the flow of air. I carefully asked around, but I didn't find a single female colleague who'd found this to be a good option. I don't have that much experience yet, but it seems to confirm that even though we can't control the direction of our pee as easily as men, the air flow does its work just fine at a few centimetres' distance.

My scrambled eggs and porridge are still where I left them, stuck to a piece of duct tape on the table in Node 1. I feel some small triumph, finding them still there. I open the packet of scrambled eggs with small scissors, an essential cutlery tool in space, along with a long-handled spoon to reach down into larger packets. I can't resist the temptation to set some of my breakfast afloat so I can catch it by taking bites in mid-air. It's bizarre how everything up here is strange and familiar at the same time. I know the name of every place and every object. I can look in any direction and see, in my mind, behind the bulkhead, the layout of the ventilation ducts that distribute air throughout the Station. I know that only here, in Node 1, the bulkhead is salmon-coloured, and that the two instruments for measuring combustion products, attached here with Velcro, are exactly where they should be, where any one of us would look for them if there were a fire alarm. When I want to phone Lionel, though, I have to ask Butch for help again, and he shows me how to use the VOIP phone.

Lionel has gone back to Moscow with the little group of friends and family, by now quite close-knit, who were there at the launch. When I call them from my crew quarters, they're having supper in a Caucasian restaurant. I imagine them sitting in front of a mountain of khachapuri and shashlik. Was it really only two nights ago that we were together in Baikonur? Already, it all seems like a dream: quarantine, the goodbyes, the launch. I've half a mind to float over to the Soyuz, to breathe in that unmistakable odour, touch my Sokol suit. To embed the memory before it begins to fade away. I've never been good at remembering. The past slips off me as easily as my clothes.

International Space Station, 25 November 2014

'Good morning, Houston, Huntsville, Munich, Tsukuba and Moscow! Expedition 42 is ready for the morning DPC!' Butch exclaims in his Tennessee accent, his voice full of energy and enthusiasm, initiating my first workday in space. Up here, Butch goes to bed super early, just like he does on Earth, and he gets up at five in the morning to do two hours of sports before the day starts for most people. Even amongst astronauts, who are known for being dedicated to their work, Butch's zeal stands out, fed by his uncommon religious fervour. Meticulous and precise, he made his radio call at exactly 7.05 as scheduled, not a second before and not a second later. Terry and I float nearby, in the Lab's office corner. Along with the radio panel, there's a sort of small table and lots of office supplies, from Post-it notes to pens, all of them stuck to the wall with Velcro. This is where, every working day, we'll start our activities with the morning DPC, or the Daily Planning Conference, a conference for coordinating the day's work with our control centres worldwide.

The first one to respond to Butch's call is Jeannette, an astronaut colleague who is working today in Mission Control in Houston as CAPCOM, or capsule communicator, the person deputed to talk to the Space Station on Space-to-Ground. 'The plan is on board,' Jeannette begins, implying that the schedule on the timeline viewer is up to date. She quickly covers a variety of issues, asks questions, answers outstanding queries. Butch does most of the talking, since Terry and I don't understand that much. Butch has been working with Mission Control for months, and they talk like any other group of people in the same business who periodically put their heads together for a progress update. We'll catch up gradually, thanks to the handover sessions scheduled for this first week.

'If there are no further questions, today's PAYCOM is Crissy and she's ready to talk to you from Huntsville,' Jeannette concludes.

'Thanks, Houston, we'll speak to you throughout the day. Huntsville, good morning from Expedition 42!' Butch starts talking with Crissy about activities that concern NASA's experiments. The Space Station is first of all a research platform, where PAYLOAD is jargon for experiments and PAYCOM for the name of the communicator position in Huntsville. If the ISS were a lab on Earth, we'd say that Houston manages the infrastructure, making sure that the building doesn't collapse and that there's heating and hot water, while Huntsville oversees the implementation of research activities. These roles are fused in the same control centre for Columbus, the European lab, and the Japanese one, the JEM: with EUROCOM and J-COM we discuss both systems and experiments. Finally, Butch calls Moscow and hands over to Sasha. He, Anton and Yelena take part in the conference from the Service Module. It's notable how long the Russians speak via radio, since TsUP relies less on the timeline viewer and other electronic documents for the exchange of information. For this reason, Butch deselects the Space-to-Ground channel 1, or S/G 1. It will be used for the rest of the day by the Russians to speak to Moscow, while we are left with channels S/G 2, S/G 3 and S/G 4.

For all of us, the first activity of the day is a review of emergency procedures. We tour the Space Station together, looking over the equipment: the ventilation shut-off valves, the large electric torches secured to every hatch, portable pressure gauges, fire extinguishers, emergency masks. When we finish our inspection, we review procedures and the role assignments – the ones we established in Building 9 during the drills last summer, in a terrestrial life that already seems very far away. Maybe it's normal that our arrival at the Space Station only yesterday morning already seems like a distant memory. I'm constantly learning, absorbing images and sensations, finding new ways to move my body in space. My perception of time seems to be expanding.

At the same time, I'm frequently reminded by life on the

Space Station that I'm a rookie. I continue to move around awk-wardly, often setting myself up for a collision. While I'm float-ing in mid-air, it's impossible to correct myself before I bash into something. Moving in the Lab is particularly difficult, because there's not much free space on the walls. The equipment is set up everywhere, port, starboard, deck or overhead: thick cables, large connectors, drawers with protruding handles, laptops – each with a partially coiled cable and bulky converters. We also keep frequently used items in the cabin, just attached with some Velcro. Despite the apparent disorder, I know that everything is exactly where it should be – at least until I get there, all thumbs, and send things flying.

I'm also not comfortable yet with many simple everyday tasks. One colleague described the first week on board as an attempt to drink from a fire hose. I'd say the description is fitting. I'm perfect-ly trained on every single aspect of working on board, but only experience is going to teach me how to translate that knowledge into efficient execution. For now, I continue to bombard Butch with questions, and luckily he has endless patience.

International Space Station, 27 November 2014

No DPC today. My fourth day in space coincides with Thanks-giving Day, an American holiday that's celebrated on board. I've always liked the idea of dedicating one full day in the year to being thankful. For my part, I'm very grateful that my mission has started out with such a calm week. Poor Butch didn't have a real day off for a month when he first arrived.

Yesterday I performed my first experiment, 'Blind and Imag-ined'. It wasn't an easy debut, since I had to install a lot of equip-ment and laboriously calibrate four video cameras, then proceed to perform a long series of pre-planned movements that were re-corded to evaluate motor control and balance. Not at all a diffi-cult task after a few weeks in space, I imagine, but on my third day it was really challenging to manage this long and elaborate protocol. It all took much longer than planned, yet in the end we successfully captured all the data.

I hope the researchers will be able to draw useful conclusions about neurological adaptation in weightlessness that may one day help in the rehabilitation of motor and balance problems. As far as I'm concerned, it's fun to observe, however unscientifically, the games my brain is playing with me. For example, if my feet are pointing towards the ceiling or one of the walls, then my head automatically decides that that's the floor. It's probably a useful adaptation, which tends to alleviate your sense of disorientation, but I have the impression that it also interferes with my mental map of the Space Station's topography: when I need to turn into a side module, I often find myself going the wrong way, towards the JEM instead of Columbus, for example. Butch assures me that it will change in a few weeks, when I'm more familiar with the Station, and my brain learns to grasp visual references quickly. Anton, on the other hand, pokes fun at me: so why can't I just stay on the floor? Well, maybe it's because I've just arrived, but I'm having fun inhabiting space in its three dimensions. I can spend the rest of my life on the floor. I often eat floating at the ceiling, with my head down, like a bat. Anton maintains I'll lose this vice, too, in a few weeks. We'll see. I think Anton's perspective is influenced by the internal organization of the Service Module, where he spends most of his time, and there's a clear distinction between up and down, as if it were a terrestrial environment. The Russians would never think of putting the water dispenser on the ceiling, the crew quarters on the floor, a treadmill on a wall. But I love them like that.

Terry and I start right in today with our workouts, a daily requirement that we'll continue right up to our re-entry. Without constant training, weightlessness causes not only muscular atrophy and cardiovascular deconditioning but also a significant loss of bone density, a sort of accelerated osteoporosis. We're lucky to have extremely efficient gym equipment on board the ISS. The simplest machine, and the one I grapple with first, is the CEVIS: Cycle Ergometer with Vibration Isolation and Stabilization. It's a stationary bike, though anyone looking for a seat will be disappointed: it makes no sense to sit down in weightlessness. So I put on my cycling shoes – the first shoes I've worn since the launch – attach my feet to the pedals, and tie a strap around my waist. CEVIS reveals itself to be as tough as I expected. No one was able to explain it to me exactly, but its mechanics are notoriously different to those of a bike on Earth, so much so that it's hard to follow your usual training protocols. To help with my pedalling, I push myself down with my arms, holding on to the metal frame of the CEVIS. It's all fixed on a wall rack, but with a trick: a vibration isolation system dampens the loads transmitted into the ISS structure. In the long run, this extends the structural life of the Station, and in the immediate term it meets the demands of the scientists, who definitely wouldn't want vibrations, kicks or jolts to disturb their experiments. It is, of course, impossible to restrict all impact to the structure, and small disturbances do take place anyway for the simple reason that the Space Station is not an abstract mathematical point, but a body with a finite extension. That's why, to be precise, we don't speak of weightlessness but rather of microgravity. Of course, I can't actually feel the difference. The small residual effects of gravity are registered only on sensitive accelerometers scattered throughout the Space Station.

Butch is our host for the Thanksgiving Day festivities. In Node 1, he's prepared a feast of caramelized sweet potatoes, cornbread, mushrooms and, of course, smoked turkey. For dessert, there's a cherry and blueberry cobbler. All of these are standard ISS food packets that Butch set aside during various on-board menu rotations, some of them freeze-dried, others ready-to-eat after a spell in the little oven. We were all amused to read out some of the more extravagant cooking suggestions from our Daily Summary, a document Houston sends every day to update the crew on the Space Station systems, answer outstanding questions and communicate the names of staff on duty in the various control centres. It ends with this phrase: 'End of operational content'; then there are usually a few pages of cartoons and jokes the control centres hope will start our day with some hearty laughter. They've gone wild today, and Riccardo, the friend who put the finishing touches to our Soyuz logo, has managed to put in a made-up Thanksgiving menu from Milliways, the Restaurant at the End of the Universe that lends its name to the second book in Douglas Adams' Trilogy in Five Parts.

We've done our part, too, to entertain our control-centre colleagues. In an honest match on Space-to-Ground between Butch, Terry and me, Butch was justly deemed to be the undisputed winner of the contest for the Best Turkey Imitator in Low Earth Orbit.

International Space Station, 29 November 2014

'Always remember that we're in a fish bowl up here,' was one of Butch's first warnings after our arrival. At first I wasn't sure what he meant.

Was it because we were floating around like fish in water? No: it's because the walls are like glass, and we're always under observation. Every day, as soon as the morning DPC is finished, a wide-angle camera begins transmitting images from the Lab and Node 1 to Earth. Another one transmits images from JEM and, in the background, Columbus. There are cameras in Nodes 2 and 3 too, but since those areas are for living, hygiene and sports, the cameras are usually turned to the wall unless there's some particular need or request. Today, however, the cameras are all off, or at least not transmitting anything to Earth. It's the beginning of a quiet weekend in space.

I get up late, feeling rested and fit. Thankfully, I no longer feel that pressure in my head and I've stopped putting a sickbag in my pocket just in case. I get out of my sleeping bag and cross the Lab in my pyjamas. I'm beginning to manage my floating better: I still run into things every now and then, but not as often and in a more controlled way. I'm learning to be more precise with my pushes and I use only the force required, which is actually not that much. I remember how Cady tried a few years ago to describe weightlessness to an actress who was preparing to play an astronaut in space. 'Imagine that you're holding a strand of hair stretched between your fingers, and you're using it to push yourself along, with a touch so light that the hair won't break. That's all you need to get yourself slowly floating towards the other part of a module.'

In the pantry, the box with the breakfast packets is almost empty, so I dive into the PMM to look for another pack. The Pressurized Multipurpose Module, which is the main stowage place of the ISS, was supplied to NASA by the Italian Space Agency. Butch has also ingeniously turned it into a 'shower room' by installing over the entrance a makeshift curtain that can be closed whenever necessary. A kindness towards me, maybe, but perhaps also so that he'll feel comfortable. The PMM is a great stowage module, but it's pretty cold and doesn't have proper handrails. Sooner or later I'll have to tell Butch that I prefer to wash with a bit less privacy in Node 3: it's more comfortable and welcoming. Sure, it's also very busy in there, at least during work hours, and everyone has their own views on how much flesh they want to see, especially when it comes to the opposite sex. I'll bring it up later, when we've got a bit more used to each other. For now, it's the PMM.

'*Dobry den*, good morning, Samantha.' Anton is suddenly behind me, floating in from the Russian segment in his usual good mood. We don't see much of each other here on the Space Station. The cosmonauts work almost exclusively in the Russian segment, and we go there only rarely apart from evenings, when we often eat together and chat for a bit. After having spent weeks together in Baikonur and shared the intense experience of the launch, it's strange to see so little of Anton. Now he's come to bring back my spoon. Again. It flew off once already this week, inevitably heading for the Russian segment because of the way the air circulates, through the ducts towards the American modules and returning to the Russian modules through the open hatches. My spoon, like all lost items, floated with the airflow towards the first ventilation grid, in this case the one in the near-by FGB module, where Anton found it while cleaning. He hands me the precious utensil and goes back to his vacuum cleaner, floating nimbly in the funnelled tunnel that connects the American segment with the Russian one.

Saturday is housekeeping day for the whole of the Space Station, and I, too, will get busy with a vacuum cleaner later on. There's nothing else scheduled for today, other than our weekly video conference with our families. The calls are not so different from common video calls on Earth, except that the audio goes through the Space-to-Ground radio on a privatized channel, configured so that no one on Earth can listen in. On board, it's simply considered good space etiquette to avoid listening to others' conversations.

So when I go through to the Cupola, I hurry to the radio panel and deselect channel SG/4, since Sasha is using it in another module to talk to his family. It will be my turn in about ten minutes, and I'd like to have my first video call with Lionel here. I log on to the computer and wait, enjoying the last of the dying sunset. Then suddenly I see it, far off towards the north: a green tongue of flame snaking along the horizon in a hypnotic dance, like the graceful, shimmering veil of a ballerina, fluttering with her sinuous moves.

It's my first aurora borealis.

31.

> This story seems insignificant. In truth, it was a great adventure. I saw lucidly in that room what I could never have discovered on a visit to the farm as a tourist.
>
> Antoine de Saint-Exupéry, *Flight to Arras*

International Space Station, 2 December 2014

Up here, when I wake up, I'm already at work: all I have to do is stretch out my arm and open up my laptop. There are about twenty of these Station support computers, as they're called, distributed throughout the modules. They're normal PCs running on Windows, and we use them to look at the timeline viewer, procedures, stowage notes, to send and receive email, make phone calls, manage our photos – in short, for everything that isn't strictly command and control of the Space Station. There's a second laptop in my crew quarters, isolated from the on-board network, and it allows a remote connection to a computer in Houston to access the internet. It often has issues, and when it does work, it's agonizingly slow. When it's at its worst, it can take ten minutes to write a 140-character tweet. I'll type a few words and then I'll write an email or read a document on the other computer, while the characters in the tweet appear in groups of three or four letters.

This morning, the timeline viewer tells me not to eat, drink or brush my teeth before I collect samples of my saliva. I prepared the swabs yesterday and I'm very familiar with the procedure, since I did this several times on Earth, each time for one-week periods. I put the first swab in my mouth and open the door of my crew quarter to signal that I'm up. Sasha lives across from me, in the overhead crew quarter. His door is closed, so he could be in the Russian segment already, or else still sleeping. There's more than an hour to go before the DPC. Terry lives in the starboard crew quarter near my feet. When the doors are open, we could be roommates talking to each other from bed, or rather, our sleeping bags. It's a bit like being in a university dorm.

'Mmmmm,' I say to his 'Good morning,' pointing to my mouth and miming the number four with my fingers. The first swab has to stay under my tongue for four minutes.

'Ah, OK.' We don't all do the same experiments. Terry's not taking part in this one, which is looking at how disruptions to astronauts' immune systems renders them more susceptible to infection.

Some of the activities on the timeline viewer are written in pink to indicate that the schedule is flexible: I can do them some other time if I want to, as long as I don't get in anyone else's way. Sport is a good example. It's always pink, and there are two and a half hours of it, split between cardio training on the cycle or treadmill, and resistance training on ARED, a weight-lifting machine. Easy, you'll say, to lift weights in weightlessness. There's a catch, though, and you can see it in two large vacuum cylinders that provide an adjustable load, up to 300 kg for some types of exercise. This makes ARED an efficient countermeasure, allowing you to compensate for the effects of a life with no significant physical effort. So once a day, I grab my gym shoes and set up ARED for my training session. My routine always begins with a few variations on the squat, perhaps the most important exercise for limiting the loss of bone mass in critical areas such as your hips and pelvis. When you perform a squat on ARED, not only does the bar move up, but the platform you're standing on moves downwards. In order to protect the ISS and any experiments from vibration, the entire machine is free to oscillate and slides on special tracks. The result is an undulating movement which can be somewhat disorientating, so much so that at the beginning I often had a sensation of falling forwards. Now I'm much more used to it, and only somewhat worried that one of the cables for load transfer has been broken a lot recently. I watch it constantly to make sure it's sliding freely in its track, and when I'm finished, I look to make sure there aren't any damaged fibres. We're using up the spare cables and won't receive any others for several weeks.

I'm still very inexperienced and I often struggle to keep up with the red line that runs along our timeline viewer, marking the inexorable passage of time. Luckily, Butch is always ready to help. I usually don't need much, just some reassurance that I'm doing the right thing. One of his 'Looks good!' saves me from checking the procedure again and allows me to go on to the next thing confidently. I'm constantly thanking him. 'You bet,' he says as he turns to float back to his work. 'We all need confirmation cues at the beginning.' There it is: confirmation. Confirmation that my brain is working well, that I'm not about to cut power to an experiment by removing the wrong plug, for example. I feel more insecure up here. I don't trust my memory, and I'll check a number I've just read three times. Can it be true what they say, that we're all a bit stupid in space? Yet my WinSCAT the first of a series of monthly tests I'll take on board for objective measurements of various cognitive abilities, gave results more or less similar to the pre-flight ones. It registered only slightly slower on my reaction time, but I think that's to do with my having been floating in front of the keyboard rather than sitting solidly on a chair. So why all the uncertainty? Is it because of the huge responsibility I feel on my shoulders? Mostly likely it's just temporary overstimulation. My brain is learning how to manage geometry and movement in a situation no human or living being on Earth ever experienced before the space age. Neither millions of years of evolution nor thirty-seven years of life have prepared me for weightlessness. All the same, a half century of spaceflights has taught us that humans are adaptable. I just have to be patient. While my brain is busy changing me into a space creature without my having to think much about it, I have to accept it's being a little less prompt and timely in remembering a number I've just read – and reread it as a precaution, because now this extraordinary human outpost in space is also in my hands.

Float here and there, and the day passes in a second. After the evening DPC I go and rummage through my things in search of a large t-shirt with 'The Answer Is 42' written on it. It's my birthday gift for Terry, who's turning forty-seven today. It's always good to have an extra t-shirt, I thought, and this breaks the monotony of the single-colour cotton t-shirts that make up part of our weekly workday livery. We're going to celebrate in the Service Module, which has become our gathering place for crew dinners. Its snug, homely atmosphere, the yellow fabric on all the walls and the simple fact of being able to float around a table makes the Service Module the perfect place to spend good times together.

International Space Station, 3 December 2014

Along with my saliva samples, today I'll be giving science five phials of my blood. There's in fact a wide range of blood biological markers that are very useful in understanding the process of adaptation to weightlessness. Terry has been asked to assist me and he is floating, still drowsy, in Columbus, where we keep the blood draw supplies. There's a large square bag on the floor full of equipment for which there's no room inside the racks. I slip my legs under the bungees that are anchoring it to the floor and then move them up over my quadriceps in such a way that I can almost stay 'seated'. After a few clumsy attempts, we succeed in getting our first sample; it's lucky that I have two arms! Terry and I organize the phials on a piece of duct tape and float off to breakfast and the DPC.

The phials have to coagulate for half an hour and then spend another half hour in the centrifuge before they are ready for the freezer. There are three freezers on board, and keeping them at -94° C takes plenty of energy, so there are very strict time limitations on opening them. Wearing very thick gloves to protect your hands from the cold, you open one of the long trays, and with the immediate release of a cloud of condensation, the sixty-second countdown begins. The procedure tells you exactly where to put the samples: which freezer, which dewar, which tray and which tray section. My biggest fear is that I'll make some mistake with these tiny details. A sample in the wrong place, and then someone – it could even be me – will have to look for it eventually, and that's if there's even time before the designated re-entry vehicle leaves. Or maybe it will only be someone on Earth who realizes that a sample is missing. Result: loss of science. The three words no one wants to hear.

How are you sleeping? How's your appetite? Digestion? Have you taken any medicines since the last time we spoke? I really have nothing to complain about – I'm feeling really well – but Brigitte bombards me with questions. It's her job. We see each other once a week for the so-called 'private medical conference': fifteen minutes to talk about any possible health problems, but also about the workload and anything else an astronaut wants to discuss with someone who's obliged to keep the information confidential, unless of course there's a problem that would impact the mission. In that case, the flight director would be notified immediately.

Brigitte also asks whether I've noticed any problems with my eyes. No, and I haven't mentioned anything to worry her, but the question doesn't surprise me. For several years now, they've been paying particular attention to our eyes on the Space Station, since it's been shown that many astronauts experience a loss of visual acuity. At present no one knows why this happens, though there are various theories to do with greater cranial pressure and maybe a genetic predisposition, which would explain why the problem affects some astronauts and not others. To discover the cause, we'll check our eyes with a variety of instruments, from the ophthalmoscope, which takes photos of the retina, to optical coherence tomography, which makes a 3D eye scan.

Terry and I are helping each other out today with the eye ultrasound, aided by a specialist on Earth who instructs us remotely on how to position the probe. I've discovered that in space we don't need the gel typically used in ultrasounds to facilitate wave transmission: we can just use water. Terry lets a little out of a packet and the bubble it forms floats in midair. I approach it, until it touches my eye. As long as I move smoothly, the water stays there, as a gelatinous mass would do on Earth.

This evening I get to the Cupola in time to see us fly over Italy. At last! I spy the alpine valleys of Trentino, where I grew up, while Milan, the city where I was born, is a big yellow mark blurred by the thin layer of clouds over the Po Valley. The rest of the boot is neatly outlined by a filigree of light: if the Earth were an elegant lady in an evening gown, Italy would be her gaudiest jewel. My gaze embraces the whole of the country. There are the places I've lived in: Naples, Lecce, Foggia, Treviso; regions I've flown over in my jet at 300 m in altitude and over 700 kilometres per hour. Now I'm gliding along the Appenines at 28,000 kilometres per hour, but I have no sense of speed. Up here, there's no landscape nearby to speed past – only oceans and faraway continents, which unfurl beneath me languidly and majestically.

International Space Station, 8 December 2014

Drinking coffee in the morning is one of life's pleasures, both in space and on Earth. Before the morning DPC, I fill my packet with water from the dispenser and take it along, together with my tools, when I float towards my first activity of the day. There's a running joke on the Space Station to the effect that yesterday's coffee becomes tomorrow's coffee, a light-hearted reference to the water recycling process. And it's true: even urine gets this treatment, thanks to the complex equipment in the rack under the toilet. The non-recyclable residue ends up in a large cylindrical canister, which is periodically emptied into one of three tanks on ATV that arrive on the ISS with water for the Russian cisterns. It's strictly for the Russian ones, since their water is sterilized with silver ions, whereas NASA's water contains iodine. There is always some danger of emptying the urine canister into the wrong tank, one containing potable water not yet transferred on board. That's why I always check at least three times to make sure I've connected the transfer hose in the right spot. The canister is one of the most massive things you have to handle on board. It weighs nothing up here, but it retains its mass and inertial properties. In other words, if you gave it a shove to make it float across the FGB module, which is long and narrow like a tunnel, it would tend to maintain a stable direction and I could let myself be pulled along, like a witch on her broom. There's a danger that this could be highly amusing. Hypothetically speaking, of course.

I have to admit to some guilt, and confess that I am contributing nothing to today's water recycling. All in a good cause, however: I've given my pee to science. I'm taking part in various experiments that require me to take urine samples over a twenty-four-hour period. It's one of the things I was a bit nervous about, since the logistics of collecting urine in weightlessness are not as straightforward as filling a plastic cup on Earth. I even tried it out in Baikonur a few days before we left, using some collection bags I'd been given in Houston for practice. Luckily they work really well, both down there and up here in space.

This evening, I finally called Mary and Stacey to say hello. The strange thing is that, because they live in Houston, a call from space is like a local phone call. After many failed attempts, I realized that the dialling code I was using wasn't necessary; you needed it only for calls outside the United States. I was, in fact, using the international dialling code. How strange to think locally from up here! Even stranger to hear a voice saying mechanically, 'What's your emergency?' In my attempts to reach Mary and Stacey, I omitted one of the zeros and ended up calling 9-1-1.

International Space Station, 10 December 2014

I can think of nothing worse than having to chase after liquids or solids floating away from you while you're using the loo, and nothing more embarrassing than having to confess to your own crew that they might experience a particularly unpleasant close encounter. Tales of legendary mishaps of this sort have circulated, and I'm guessing that some of them may even be true. When I heard Butch go into the bathroom right after me, and then shout 'The toilet insert!', along with one of his polite fake swear words, I knew I'd made the second most embarrassing mistake. Above the toilet's solid-waste container there's a small seat with a lid, which forms a sort of miniature w.c. Since you don't really sit down, almost everyone finds it easier to lift the seat up, and push the buttocks directly over the circular hole underneath, the opening to a short tube that goes into the container. At the end of the tube, there's a semi-rigid curtain, which allows you to push something into the container and prevents it from coming back out. Last but not least, there's a single-use insert, a small bag made of thick, clear plastic punched with numerous holes that allow the air to suck things in. Besides leaving the bathroom 'clean enough to eat on', as one Space Shuttle commander is reported to have said, space etiquette requires you to put in place a new insert for the next user before you leave the toilet. I hadn't done that, and Butch didn't notice in time.

'I'm sorry, Butch!' I shout through the toilet door, hoping he can hear me above the noise of the fan.

Happily, Butch recovers quickly from his initial surprise. 'Oh, don't worry. I have two little girls, so I'm used to putting my hands in poo.' It's true what they say: parenthood teaches you everything.

I wait for Butch to emerge so I can apologize again, and then I return to the sample collection I'm working on. The experiment is called Microbiome and it's a study of the non-human inhabitants of the ISS. I'm not talking about aliens, which must be well hidden if they're here, but the billions of microorganisms that share our space up here, many of which live in our intestines and impact our state of health. Other than taking samples of blood, urine and faeces, Microbiome requires that we swab the surfaces of various components that get touched frequently, such as the shutter knobs in the Cupola, for example. While I'm in there, I sneak a peek out of the window. We're over an expanse of water, and the map shows that we're about to fly over Patagonia. It's one of my favourite places, with the snow-capped Andes and lakes nestled in the valleys, their water painted an intense turquoise by the fine glacial sediment. I take a couple of photos, disappointed that the southern tip, the famous Tierra del Fuego, is covered by clouds again. I'll have to look out for it another time – I still have five months left.

International Space Station, 13 December 2014

The radio is quiet on weekends, for hours at a time. I'm tempted to say that silence reigns: I no longer hear the continuous hum of the dozens of pumps and fans that are left on. I only notice them when I realize how impossible it is to talk from one module to another. There is an advantage, in some ways: the continuous hum covers voices so that you can actually have a private conversation with your family or the flight surgeon without necessarily shutting yourself up in your crew quarters. To escape passing traffic, we often use one of the computers in the side modules, such as the JEM or Columbus. My favourite spot is in the JEM, not far from the two round windows in the endcone, and I often take my video calls in there with my feet under a bungee on a wall, doing a yo-yo – with gentle nudges, of course. More vigorous ones would cause Houston some alarm, as it's monitoring the Space Station's acceleration. I'd get a call for sure, even on Saturday. You don't play games with microgravity on ISS.

Every now and again, we call on Saturdays to get remote help with our cleaning. Before removing the panels and running the vacuum over filters in a given module, we ask Houston to turn off the smoke detectors in order to prevent false alarms. We could do it ourselves, but the controllers willingly give us a hand. Working on Saturday must be pretty boring for them. CAPCOM usually calls a few minutes later to confirm that things are switched off, announcing, 'The crew is prime for detecting smoke.' In other words: make sure there's at least one human nose around until the detectors are turned back on.

Although it is Saturday, we have work to do, since Dragon will arrive in just a few days. For me, this is an important moment: for the first time after all the years of simulation I am at the controls of the real robotic arm, the SSRMS. Right now, it is a powerful and imposing presence outside the windows of the Cupola. I've lost track of how many pictures I've taken. It's so beautiful out there, outlined against the horizon. My job is to move the end effector of the SSRMS from a distance of 5 m to the grapple fixture designated for this exercise, and to stop the moment when I would press the trigger to initiate capture. Butch takes the controls for a few seconds and, returning a favour I've done him earlier, he dutifully misaligns the SSRMS relative to the grapple fixture while I look away so I won't see the movement his hands make on the commands. It would make it too easy: I have to figure out which corrections are necessary from the alignment cues on the target, which is drawn around the grapple pin, and I have to make the corrections in motion as I approach. Butch says he's satisfied with the starting position and cedes the controls.

I'm a bit nervous. This is not a simulation. There are 14 metres of robotic arm moving outside. It's one of the most critical components on the Space Station, essential for grappling cargo vehicles and for many of the repairs in EVA. As soon as I deflect the manual controllers, I find to my relief that the SSRMS is a well-domesticated beast: it moves tamely, following my commands, and is much more stable than the simulator arm, which accustomed me to managing much more significant oscillations. Train hard, fight easy.

International Space Station, 18 December 2014

'The plan is <u>not</u> on board,' CAPCOM told us at last night's DPC. What she meant was that we shouldn't trust the timeline viewer, because it would be modified overnight. Dragon's launch, set for tomorrow, has been put back to January, and all our activities now have to be rescheduled. What a shame! All my Christmas gifts for the crew were already in the bowels of Dragon, in my Crew Care Package. I'll have to explain that Father Christmas ran into some difficulties.

I spent the greater part of the morning doing repair work on a piece of failed hardware, and now it's time for a run. T2 isn't very different to a common treadmill on Earth, apart from two things. The first is that it's fixed to a wall, on a vibration isolation system. The second is the harness, with two rings at hip level to attach one end of two variable-length chains. The other end is fixed to the frame of the treadmill, just on the side of the running belt. These chains stop me from floating away at my first step, but they also include an elastic segment that makes running possible. By adjusting the length of the chains, I can regulate the force with which they press me onto the running belt. At the moment I'm about 60 per cent of my normal body weight, and the harness is there to distribute that load between my hips and my shoulders, based on the same principles of a good backpack. In fact, I find myself shifting the load quite often, alternately relieving my hips or my shoulders. After three weeks of training, I think I've come to a final judgement: I do not get on with the T2 harness. This strange kind of running, however, on legs that don't weigh anything, is extremely important not only for cardiovascular conditioning, but also for maintaining bone density. With every impact of your foot on the platform, a regulating signal is sent to the bone-building cells, the osteoblasts, and this helps to maintain the delicate balance between production and destruction of bone tissue. So I run faithfully, four or five times a week, distracting myself from the bothersome harness with an episode of *Battlestar Galactica*, a series I requested be sent to me on board, one episode a day. An excellent choice!

After my run, I decide to treat myself to a new pair of trousers. I have six pairs of tactical pants – very practical – for the whole of the mission, in either khaki or army green. They're waterproof and have lots of pockets to which NASA applied strips of Velcro so they can be kept closed. It's incredible how easy it is to lose things up here. All it takes is for a pocket to open up or a moment of distraction – and that moment may later cost you a lengthy search. The other day I was looking for a torque wrench. It wasn't that small, but after a while I suddenly realized it was right there, floating half a metre in front of me. My eyes were focusing at a greater distance, on the walls a few metres further away: I'm not used to looking for things in midair.

Whenever I can, I duck into the Cupola, sometimes just for a quick look and sometimes to snap photos. The computer map shows our position and our ground track, but I like to try and guess. Naturally I know my way around my small, beloved Europe – which we overfly in about ten minutes – and I easily find my bearings when my gaze settles on lakes, islands or peninsulas with characteristic shapes. But sometimes I try in vain to find known reference points on an immense expanse of earth or sea. All the same, the Earth is becoming more familiar to me with every passing day, and I greet its colours, textures and patterns like old friends when they show up on the horizon. How could you fail to recognize the blue of the Caribbean Sea, with its infinite gradations of indigo, emerald, turquoise and cobalt? Or those cathedrals of rock and snow, the Himalayas, with their frozen lakes? Or the soft brown undulating texture of the Namibian desert? Some orbits fly almost entirely over the oceans. Then I spend a long time looking at the horizon while it spews up sea and clouds, sometimes of a creamy consistency, other times light as a dusting of icing sugar, at the mercy of the winds as they create straight roads or swirling spirals, clear faultlines or smoky transitions in sharp chips or soft white puffs. Atolls peep out between the clouds, remote and fabled, delicate and evanescent, and perhaps never trodden by human feet.

International Space Station, 22 December 2014

Activities on the Space Station follow each other in quick succession, and as long as they're not accompanied by an emergency alarm, they are all carried out with equal meticulousness and attention, but in an unruffled manner. This goes also for in-flight calls with important individuals, the ones we call 'VIP events'. On Earth, these kinds of events are preceded by days of frenzy on the part of everyone involved. Up here, however, all we get is a short, written briefing, and we'll find a block of ten minutes on our timeline viewer outlined in blue to indicate that that is a fixed activity that must happen at a precise time. It's no different from making an in-flight call with a school, for example. It's how it went earlier today for the traditional Christmas wishes that the President of the Republic addressed to all Italian military serving abroad – plus me, in low Earth orbit. I was touched when the President, perhaps a little emotional himself, dropped my formal title and told me, 'I'm calling you Samantha, because you're Samantha for all of us Italians.' The call, which was only audio for me, ended as usual after only a few minutes. I had just enough time to restow the camera, microphone and flag, and float away to get the tools for my next task. That's how it is on the Space Station: each activity is like a soap bubble. Once it's had its time, it disappears without a trace, making way for the next one.

And yet, I think back to those words, spoken by someone who is certainly not inclined, by role or by culture, to easy hyperbole. I'm not immune to vanity and I'm certainly gratified by so much affection. But for the first time, I find myself dreading what's to come, my re-entry on Earth and life after the space mission, which I've never wanted to think about much. Will I be recognized on the street? Will my country be the place on Earth where I'm no longer anonymous, where I'm the subject of gossip, where I'm spontaneously offered unsolicited favours, perhaps in exchange for a selfie? Will the price of my space voyage be a life of minor celebrity, and will I be continually unsettled by never-ending accolades, propositions and requests, as if it weren't already hard enough, even without the traps of celebrity, to lead a centred, virtuous life respectful of others? 'Celebrity is an illness that always leaves a scar,' wrote Oriana Fallaci at the time of the Apollo missions.*

At any rate, no use thinking about that now. We've got huge sacks of rubbish to take to ATV, and Butch is already waiting for me.

* Oriana Fallaci, *If the Sun Dies*.

32.

Two things fill the mind with ever new and increasing admiration and awe, the oftener and the more steadily we reflect
on them: the starry heavens above and the moral law within.
   Immanuel Kant, *The Critique of Practical Reason*

International Space Station, 31 December 2014

The clocks on the Space Station are set to UTC time, or solar
time at meridian zero, and when they say 9.00 p.m., it's time
to celebrate the first New Year, the one in Moscow. As soon as
we've all exchanged good wishes on board, Sasha, Yelena and
Anton begin calling their families. It's not an easy time right now
in Russia, and for the past few weeks many have been worried
about their savings. So all the more reason to be happy about a
New Year's wish from space! On the groaning table in the Service Module, a computer sits quietly showing *The Irony of Fate*, a
romantic comedy from the Soviet era. This warm and reassuring
film is shown on TV in Russia on the last day of every year, and
we all know it, to the extent that we interrupt our conversations
at certain points just to watch the best scenes. When it's over, we
bond in a sort of karaoke session with songs from the film inter-
spersed with some from Adriano Celentano, and a bit of improv
from anyone who feels like singing. At one point Sasha even goes
off to get the guitar. We're having the most discreet party on or
off the planet: however loud we are, no one can hear us outside
our metal shell. We're enveloped in the hermetic silence of space.
   Who knows how perplexed alien visitors might be if they
passed near our solar system with fancy technologies worthy of
a science-fiction film and, their curiosity piqued about our little

planet, almost entirely covered by a thin layer of water, sent a group here to look around on New Year's Eve itself – a moment with no particular significance, astronomically speaking. I can imagine them going back to their mothership, with an amazing propulsion system faster than the speed of light, and reporting: 'The planet is dominated by microorganisms that have obviously had great success in nurturing a wide variety of complex species to satisfy their needs for mobility, nutrition and reproduction. The most notable of these species is a biped with almost no fur, found all over dry land. Every hour, the bipeds located in a specific longitudinal range abandon themselves to rituals of uncertain significance during which they consume ethyl alcohol and shoot into the sky generators of light, smoke and noise – a practice apparently unconnected to war. One particularly bizarre group consists of six bipeds locked in a metal container about 400 kilometres from the Earth. The absence of a system for generating artificial gravity might suggest that this is a structure for punitive detention. It could also be a means of isolation to protect others from contagious illness or other dangers. We have judged it prudent to avoid all contact with them.'

Actually, we're not that isolated up here. Besides the usual Sunday video calls to our families, there was one on Christmas Day and there will be another one tomorrow. I've even been able to talk to my loved ones while we were flying over them, and I imagined our eyes meeting somewhere halfway between Earth and space. Lionel hung the Christmas tree from the ceiling in our apartment, imitating the one Butch put up a little while ago in one corner of the Lab, together with six stockings hung neatly over the hatch – one for each of us. During the past few weeks, whenever one of us has come across someone else's favourite packet of food, we've put it into their stocking. Yelena adores cappuccino, for example, while Anton especially loves NASA's meat dishes. I also put into everyone's stocking some packets of my bonus food. My actual Christmas gifts of chocolates and sweets are still stuck down on Earth. They may turn out to be useful for our next Christmas in space – the Orthodox one. It will definitely be hard to repay Sasha, Yelena and Anton for the large beautiful foulard they gave me. Unfolded, it moved sinuously in the cabin like the legendary flying carpet. In that festive atmosphere, Butch gave me my golden Astronaut Pin. In accordance with NASA tradition, by making it to space I had earned the right to wear it.

When the clocks on board say eleven o'clock, we celebrate the arrival of 2015 on the European continent, and imagine the fireworks displays in Cologne, Rome and Paris. We then wait for the stroke of midnight on the Space Station, singing tunelessly and chatting away. There's a buzzy cheer. Anton and I engage in a silly dance over one of the windows in the Service Module, hoping to greet the new year by dancing on top of the world. Then everyone starts to feel tired and not up for waiting till six in the morning, when Houston will welcome 2015, so we gradually break up for the night. I hug Yelena and Anton and, with the last 'Happy New Year!', I float away with my roommates to Node 2. I'm no longer clumsy, having learned to give just the right push to move smoothly through the Space Station. I feel at home up here now and, however hard I try, I can't really imagine what it was like to walk, to feel my own weight or the pavement resisting my feet. My feet are losing their calluses. Soon they'll be as soft as a newborn's.

Before I get to my crew quarters, I treat myself to a first glimpse from the Cupola in this new year. Now that I'm more at ease with life and work on board the Space Station, now that I don't struggle to keep up with the timeline, now that the postponement of the Dragon mission and the arrival of Christmas have thinned out our agenda, I have a bit more time to contemplate the beauty of our planet. Sometimes I stay in the Cupola for an entire orbit before going to bed. As the Space Station wraps the Earth in its embrace, I watch a silent film unfold before my eyes with an endless flow of seas and continents, mountains and deserts, glaciers and lakes, forests and cities. I can see for around 2,000 kilometres in all directions; further away, the curvature of the planet conceals oceans and lands behind the horizon. This view is restricted and vast all at once. Vast for human eyes, used as they are to being confined to the surface of the Earth, but restricted when compared to the dimensions of the planet. If the Earth were the size of a billiard ball, we would be little more than a few millimetres away, and could see less than a coin-sized patch of its surface.

I can't deny it: the orbits I love most are those that fly over Europe. Sometimes we arrive from the northwest, and once we've crossed the ocean, we sight land first on Ireland's south-western peninsulas, those fingers sticking into the Atlantic, announcing the British Isles. We are soon over the continent, and then down towards the Mediterranean. At night it's particularly moving: London and Paris, brilliant and unmistakable, call to one another from across the Channel; from Calais to Amsterdam and right up to Cologne, it's all a dense web of warm light, and then the human presence thins out until, beyond the Alps, I make out the unique contours of Italy. The peninsula shines brightly, beautiful and still, like a sleeping princess. Sometimes, the Mediterranean is a black well; at other times it offers the sublime spectacle of the moon's reflection, in a milky light of mysterious beauty that directs the attention of the entire universe to this patch of sea, as if Venus were just about to emerge from the waters. Other times we fly a little more to the east, between Greece and Turkey. I can almost hear Homer singing as one island after the other, one tongue of earth after another comes into view, bathed in the snowy light of the Moon. A little farther on I see Cairo, and from there the Nile steals the show, snaking towards the heart of Africa like a diamond necklace on a black dress. In the distance, the stars emerge from behind a band of spectral green, verging on yellow, which wraps the earth in a diaphanous veil. At times it's wider, and sometimes it's thinner, often delineated by a cleaner, more intense line that marks the transition into the blackness of space. Occasionally, this colour display continues higher up, with brilliant strips or splashes of violet suspended above the horizon. It's the nightglow of the upper atmosphere. Practically invisible by day, when the planet appears to be wrapped only by a thin blue ring, the upper atmosphere reveals itself to the eyes at night, in a dance of photons emitted by molecules excited by solar radiation. It's a symphony of colours that resounds silently when the Sun slides below the horizon, like a music box that continues to chime after the clock winder has gone.

Today, however, I see none of this. For several days we've been flying around the Earth in a condition called 'high beta', a reference to the letter of the Greek alphabet that indicates how our orbital plane is orientated with respect to the Sun. At the moment, the angle is so big that we are constantly in the light, as in the polar summer, when illuminated by the midnight sun. These are complicated days for the Space Station, because the continuous illumination puts the thermal regulation system to the test. It was quite warm today in the Russian segment. The rest of the Station stays at the usual 22° C, more or less, though not without effort: the controllers on the ground have to accurately manage the orientation of the external radiators and take other measures to guarantee a comfortable temperature for us and the equipment. Because of these restrictions, it would be impossible to welcome Dragon right now.

For us humans living up here, the high beta means that we haven't seen a real night for days now. We're flying along the terminator, the line on Earth that separates light from darkness: before us we have the weak Sun, low on the horizon, projecting long shadows, and behind us an uncertain twilight, never dark enough to reveal the stars or city lights. When there are sunsets, they linger, like passionate, reluctant farewells. But the Sun is teasing us: it hovers just a little below the horizon, and well before darkness falls over the planet, it shows up again to flood the world with fiery orange. Ironically, it's in precisely this state, while we are prisoners of an eternal dusk, and our gaze is frustrated whether we turn it earthward or into space, that we ourselves are more easily seen from the surface of the planet, as a constantly illuminated dot.

With my physical eyes temporarily blinded, I find myself more often using my mind's eye. Every now and again, while I gaze at the long shadows and the delicate, rose-tinted light of dusk, memories bubble up in front of me. The New Year's Eve party bubbles are very few, recent and clear. I didn't really celebrate until I was over thirty years old. Before then, I spent those long nights working feverishly in my parents' hotel in a small mountain town. I had a carefree, independent childhood there, skiing in the winter and exploring the woods in the summer, dreaming up great adventures on every forest path, along every stream. Then it was middle school at the convent boarding school in Bolzano, where I studied hard and played hard, with all the energy of a twelve-year-old and an intensity heightened by the constant presence of girls my own age. I joined in every activity that was offered, and I was always begging for a few more minutes of recreation, for one more game. I played to win. In secondary school I discovered karate and fell in love with it. In order to train in the evenings, and so that I could escape from boarding school, where I was starting to feel constrained, in my third year I started to live Monday to Friday in a rented room. I rode my bicycle everywhere and dreamed of going to the States on an exchange programme. The following year, I did indeed travel with the organization AFS to the country of the Space Shuttle and *Star Trek* and came back with an even stronger desire to become an astronaut. Uncertain whether I should pursue engineering or physics at university, in my last year I transferred to another school in Trento with a greater focus on science. I continued to practise karate two or three times a week in Bolzano, returning to Trento by train late in the evening. The city centre was deserted, and I'd cross it at top speed on my bike to get back to the apartment I shared with three university students. And while my classmates were learning how to drive, I was taking scuba-diving and starting my first lessons in Russian.

International Space Station, 8 January 2015

Orthodox Christmas also came and went without any deliveries from Father Christmas – or, more appropriately, *Ded Moroz,*

Grandfather Frost, Russia's traditional bringer of gifts. We've been waiting for him for some time now, on board a Dragon. On the other hand, this latest delay has given Terry, Butch and me some free time to join our Russian colleagues in a rather particular in-flight call: Father Yoav has put together a small choir from the monastery at Sergiev Posad, and they delight us from TsUP with a short Christmas concert, live on Space-to-Ground. At one point they even sing a moving rendition of a famous Italian Christmas song, 'Tu scendi dalle stelle', with perfect Italian pronunciation.

If Dragon had arrived in December, these weeks between one Christmas and the next would have seen us whirling around frenetically. Dragon's mission lasts for a month, and in those few weeks, the loading and unloading of tonnes of material must be handled with great precision, and a great many experiments must be completed and the resulting samples sent back to Earth on Dragon itself. As things have turned out, this intense period has been delayed for a few weeks, so that it will overlap with other major events such as the departure of ATV and the spacewalks. In other words, we're in for a couple of very challenging months.

In the meantime, we've had a relatively calm holiday season, and our schedule has been full of routine jobs. Thinking back on all the inspections, I'd say we've had the Space Station serviced. We've checked the condition of the emergency equipment, for example, and we've also collected water samples and measured the speed of the air flow around the ventilation grids in order to identify any clogged filters. It was also my job to measure the noise levels on the Station over twenty-four hours. We know very well that the noisiest area is Node 3, because of all the regenerative equipment for recycling urine and removing carbon dioxide, and for producing oxygen through water electrolysis. If you add to that someone running on T2, the noise becomes so deafening that the doctors recommend wearing some sort of noise protection, such as earplugs or headphones. I remind myself to be more diligent in the new year.

Before now, I've only worn protection when I was the one running: custom headphones made for my ear canal, which really soften external noise while still allowing me to listen to music or film audio.

Music has also become part of my shower ritual, which now takes place in Node 3, in a secluded corner behind the toilet. Of course, it's not very secluded if someone wants to go to the Cupola, which is right next to it, but I usually take my shower at the end of the day, right after the evening DPC, when I won't be in the way. We say 'take a shower', but don't imagine that we have running water. You fill a bag with warm water from the dispenser, and then you put some on a sort of disposable hand towel, and rub your skin with it. On-board supplies allow us a fresh towel every other day, and the same goes for the change of underwear. All things considered, I feel clean enough, but there's no doubt that the shower up here is a rather modest pleasure. Still, it's a pleasure, so much so that I sometimes linger longer than necessary, when nobody needs to access the Cupola. I like the idea of ending the workday like this, with the scent of soap, a view of the Earth gliding past the large windows, and a few songs. I always begin with the same one, Paolo Conte singing 'Bartali'. It makes me feel really cheerful.

It's a pain, though, to wash your hair, and the results are not all that satisfying, since it's very difficult to remove the dirt and shampoo without a good rinse. That's why I'm happy to have fairly short hair, though even styles like mine bring their challenges. My colleagues with long hair can just tie it back, and they don't necessarily have to cut their hair up here. I have to cut mine periodically, and so the salon Chez Terry opened its doors in Node 3 on the first day of the new year. Terry has been duly trained for this delicate task. A few months ago I asked Alicia to schedule a training session at my hairdresser's in Houston. Terry played along. Nim, my hairdresser, immediately recognized his natural talent, and Terry took his responsibility with all due seriousness: Nim takes about twenty minutes to cut my hair, yet Terry took two and a half hours. It's true that things up here are quite complicated. I'm sure that even a professional hairdresser would find it difficult to style hair that doesn't lie flat on the head and to hoover up the strands as soon as they're cut. Besides, Terry feels the weight of his great responsibility: he's convinced that I'm some sort of rock star in Italy, and that Italian women are all going to want my style. Actually, he might become famous as the inventor of my haircut: 'It's the Terry.' I tried to explain to him how Italian women are some of the most elegant on the planet, and they'd hardly be likely to take lessons in style from someone who doesn't even carry a hairbrush in her suitcase. But Terry is a man on a mission – and I'd say the result is more than respectable, so much so that for days I've been getting emails full of compliments. I have noticed, though, that these are only coming from our mutual friends . . .

There's been a huge influx of emails over the holiday period. Not only have I been exchanging holiday greetings, I've also tried to send my family and friends photos of the personal objects they sent up here with me with the Earth's horizon in the background. I wanted to take these pictures as soon as I could, and then put the things back in the Soyuz so they wouldn't be left behind in the event of an emergency evacuation. In the first few weeks, though, I was reluctant to take them out of their sealed bag. These things are precious, both in sentimental terms and often also materially, and some are no bigger than a coin. It wouldn't take much to lose one: a moment of distraction while they float in the Cupola, a moment of looking away to change a setting on my camera, and then a ring, a small medal or a bracelet could float away and get swallowed up in a crack and disappear behind a rack for ever. Yes, possibly for ever, since there are definitely hidden corners on board that won't be seen or touched until the Space Station has had its day, and its final duty is to slip down through the atmosphere, disintegrate and burn up. Bits of it will surely land on Earth, or hopefully in the sea, since the debris is expected to fall into the ocean on re-entry. Maybe that lost ring or medal, having survived the flames and the impact, will be found among the debris. Who knows what story it might tell an explorer who dredged it up one day? And what a fabulous circular story it would be if that lost object were the piece of stromatolite Lionel gave me, a sedimentary structure deposited over the young Earth by the first and simplest forms of microbial life on our planet. Billions of years later, I'm taking a picture of it against a background of that blue atmosphere, which the stromatolite helped to enrich with oxygen.

The nights are back. The stars have appeared once more, and with them the sparkling cities, their intricate strands of light mapping out the human presence on the continents, so difficult to make out during the day. I've missed the nights so much. There is a particular optical illusion I love to indulge in when we're flying over the ocean, and the clouds are glowing with the reflected light of the Moon. When I've been in the Cupola for a while, with my head down towards the planet, for anyone looking in from outside, the Earth is my sky, a sky full of swiftly coursing clouds. At the same time, space acquires depth, and I'm immersed in a black sea, swimming with the stars. All I have to do to encourage this illusion, I've learned, is look towards the Russian segment of the Space Station, with its irregular shape and rugged features. Then the Station becomes a vessel from antiquity, worn out by the years, yet fearless, and I slip through the darkness in this vessel, mapping my way by the stars, as in ancient times. What a presence the night sky must have been for our ancestors, when any light produced by humans was faint and dim, when you navigated by the constellations, and their rising and their setting fixed the rhythms of life and work. What awe, what questions the firmament must have raised, what a sense of wonder sprang from it, taking shape in myth, poetry, philosophy – and eventually in science. Science thrives on questions. It has no answers that cannot be improved upon. With slow and patient labour, it alleviates humanity's material sufferings, affords us longer and more comfortable lives, more sophisticated and efficient ways of living together. But it doesn't eliminate the darkness. On the contrary. As light is shed in an increasingly wide patch of knowledge, the border line to the darkness surrounding us only expands. The more we learn, the bigger the mystery grows. There's nothing arrogant about science. It observes and measures, hazards explanations and predictions, makes errors and corrects itself. But it also lucidly embraces limits and uncertainty. It refuses to surrender to the inexplicable, but it also knows that the road to complete understanding may be so long that it's foolish to imagine an end to it.

For those who are tempted to presume they can explain everything or who entertain the illusion that they're in control of their own lives: the firmament is a teacher both humbling and consoling. Perhaps it is something like the relationship the religious have with God. I haven't considered myself a believer for a long time. The faith I inherited did not survive my adolescence, and I'd have a very hard time today imagining God as a person or religion as anything different from a human construct, a product of human history. This doesn't upset me: I have faith in secular ethics, and I've seen virtue and nastiness displayed more or less equally, whether people are religious or not. But it's true that I envy the religious their ease in making contact with the transcendent. Those who, like me, don't attend church, synagogue, mosque or temple should at least visit the vault of heaven as often as possible.

33

International Space Station, 14 January 2015

If Mission Control wanted to surprise us with an emergency drill, it would choose a moment like this one, so as not to compromise the activities on board. But we don't have surprise emergency drills on the Space Station, so it's only a lucky coincidence that the alarm goes off mid-morning, when we're all engaged in non-critical activities. It could have been different. We might have been scattered throughout the Space Station, a long way from one another, bent over some delicate operation. Or we might have been holding a cold bag full of scientific samples that had to be transferred quickly from Dragon to an on-board freezer – I did that yesterday for several hours. Or we might just have been on T2, harnessed, or in the toilet, having started but not concluded our business . . . Instead, Terry, Butch and I are all floating between Node 2 and the Lab. Sasha, Anton and Yelena are in the Russian segment.

When the siren goes off, I'm the first to realize that it's not the overused fire alarm, the one we've begun (oh dear!) not to take very seriously, since it has gone off several times, only to prove a false positive on the Russian smoke detectors. No, this time it's not the same old script. From a distance, I can clearly see that the third light from the left is on on the Caution and Warning panel at the back of the Lab. The much-feared third light from the left. There's an ammonia leak.

'Maaaaasks!' I cry with all my strength, while opening the closest store of oxygen masks. Terry and Butch start doing the same. We move quickly aft, and they stop in the narrow, funnel-shaped tunnel that connects Node 1 to the Russian segment. I go a bit further, as far as I need to see Sasha, Yelena and Anton, and confirm that we're all on the right side. Terry and Butch proceed to close the rear hatch of Node 1, and then they come further aft and close the next hatch. Now we're isolated in the Russian segment.

Ammonia is not used for cooling in here, so there's no danger of it leaking, but we can't just remove our protection: the air in the Russian segment could have been contaminated before we closed the hatches. Our oxygen masks will only last seven to eight minutes, so our most urgent need is to prepare the respirators, fitting them with ammonia filters. The ones for our Soyuz crew are in the MIM-1 module. I open the bags and take out the pieces, while Terry updates Houston, trying to make himself heard over the radio with the mask on. Butch is still floating at the junction between modules, where he can see both crews at work. We're about to change our masks when CAPCOM interrupts: 'Station, this is Houston. False alarm.' It seems we've had our surprise drill after all.

All the hands that were unwrapping, installing, browsing and rummaging through things suddenly stop, suspended in action like the limbs of puppets abandoned by the puppeteer. We hesitate for a moment, each implicitly seeking confirmation from the others that we've understood properly before taking off our masks. We're relieved, of course, but our reaction is measured. After all, none of us has manifested symptoms of poisoning, so we don't feel that there was any immediate risk, or that we narrowly escaped danger. And in the dash to follow procedures by memory and get into the Russian segment, perhaps none of us stopped to reflect on an unanswerable question: if the other side had really been contaminated with ammonia, would we ever have been able to reopen those hatches?

We're back in the Lab now, and there's an unnatural silence: as soon as the alarm went off, the Space Station's auto-response shut down all the fans. Ground controllers will take care of turning them back on. But also up here we can't just go back to what we were doing: the meticulous order of the Space Station – disguised by the apparent mess, the spaces overflowing with stuff – has been disrupted. Houston and Moscow want to know exactly which masks were used and therefore will now be at least partially unusable; what other equipment we employed; the status of the valves and hatches. CAPCOM also poses a question I find curious: 'Did one of you activate the manual alarm?' No, of course not. The alarm went off automatically, no doubt because of a faulty sensor. It happens. But I'm beginning to feel that things are more complicated than that. Mission Control has lingering questions. It's the middle of the night in Houston, but I imagine that in the garages of a number of wooden houses in suburbia, a number of specialists are getting into their cars, probably SUVs, to go to work. They're in no particular hurry – it's been established that it was a false alarm – but they mustn't delay, either.

For us, the day will return to normal, with a few adjustments to make up for lost time, as soon as we have completed our thorough post-alarm accounting, sending the serial numbers of used masks and cartridges. We also take a few minutes to discuss everything, as we expect our flight director will want a debriefing. So that we'll be ready to respond promptly to questions about what happened, we reconstruct the sequence of events together, and the actions we took. We don't know that it was only a dress rehearsal, until CAPCOM comes on the radio again: 'Change of plans. Ammonia leak! Execute emergency response! I repeat: ammonia leak! Execute emergency response!'

We look at each other for a split second, somewhat incredulous, and immediately start over again. This time it really seems like a drill. It's been half an hour since the first alarm. If there were really a leak, wouldn't we be feeling ill by now? If nothing else, shouldn't we be able to smell traces of the pungent and unmistakable odour in the air, the one we had to smell in training in order to make sure we'd recognize it? Then again, maybe the leak is in a peripheral module, and turning off the fans has stopped the ammonia making its way to us. In any case, we're soon sheltered in the Russian segment once more, behind two closed hatches, with only one thing to do: continue to execute the emergency response as we know it. Terry, Anton and I get busy in the MIM-1 above the orange box that contains the equipment. When the respirators are ready, we help each other to put them on, one at a time, following the procedure we repeated so many times: take a deep breath in the oxygen mask; still holding your breath, and with your eyes closed, take off the mask and put on the respirator with the hood; then, using both hands, smooth the hood from your forehead to your neck in order to remove any contaminated air inside before tying the laces around your neck. At this stage you can finally exhale and breathe a few times through the filters, which purifies the air left in the hood if necessary. At last we open our eyes.

When we've all changed to the respirators with filters, it's time to find out if there really is ammonia in the air, at least here in the Russian segment. We have a chip-based measuring system that's simple and easy to use, but we discover to our dismay that its battery is dead, a point we'll no doubt discuss at length during our debriefing. For now, we need to get the job done, if not with the electronic gauge, then with the Dräger tubes, which contain a substance that changes colour on contact with ammonia. In extreme cases, the Dräger tubes would tell us that we have to evacuate the Station. Or they might tell us that the ammonia concentration is such that we can filter it through our breathing over a reasonable time, as long as we continue to wear our respirators. There aren't any other ammonia filters on board. Today the Dräger tubes confirm that we can take off our respirators.

What is going on? Why did Houston tell us it was a false alarm, only to change their minds a few minutes later? It's obvious that the telemetry received on Earth is ambiguous. Evidently there is no sizable leak, true, but there could be a small, subtle one which is slowly rendering the Space Station inhospitable on the other side of those closed hatches, to the extent that we will be unable to go back. None of us really thinks things will come to that, yet there are lingering doubts all the same, and every now and then one of us casually blurts out something like, 'So what if we can't go back in there?' or a joke, to play it down: 'Butch, Sasha, Yelena – are you ready to go home early?' This could be one scenario: confinement in the Russian segment while Houston looks for a way to recover the rest of the Station, and an inevitable reduction in crew to three people. They might let Yelena stay, so that two Russians will remain on board, and ask me to take her place as flight engineer on her Soyuz.

It's all been a mistake, surely. No one really thinks there's an ammonia leak, right? It's only that some rather unusual combination of telemetry data was received, Houston tells us, and as yet, there's no convincing explanation for it. A low-tier computer broke down just before the alarm went off, a malfunction that probably caused the initial alert. At first it seemed clearly like a false alarm, since there wasn't any other telemetry to confirm the leak. It seemed as though they simply needed to carefully investigate the chain of events on the ground, and, in the meantime, we could go back to the scientific experiments that had arrived on Dragon. But while we were cleaning up following all the excitement, specialists from Mission Control observed an increase in cabin pressure. Where could it have come from? It was impossible to deny that an ammonia leak could, in fact, be the cause. At that stage they couldn't exclude it, and as a precaution against this eventuality they had us execute the emergency response. Better to have a day of compulsory vacation – it looks like that's how things will play out today – than six astronauts dead because Mission Control dismissed contradictory telemetry data.

Compulsory vacation, yes, but it's not really calm. It's the third time the fire alarm has gone off, and now the Caution and Warning panel on the Service Module is lit up like a Christmas tree. We diligently execute the procedure again, and this time, too, we conclude that there's no fire anywhere. The smoke detectors on the Russian segment, themselves prone to false alarms, appear to be in crisis due to the uncommon crush of humanity. On the other hand, Moscow is doing all it can to make our forced stay as pleasant as possible. In addition to a few obvious, but nonetheless important things, like having enough air to breathe, something ensured by the prompt reactivation of the Russian carbon-dioxide removal system and the equipment generating oxygen by water electrolysis, TsUP extends its hospitality to Terry, Butch and me by giving us permission to open up some food containers early. We are castaways on our own spaceship.

There's little we can do on board, but things must be frenetic in Houston. External cooling loop B has been isolated as a precaution, and the pump has been turned off. Luckily they haven't considered it necessary to vent the ammonia, something that would have compromised the Space Station for a long time, but they nevertheless had to turn off a lot of equipment that was in danger of overheating. The specialists are now reconfiguring the systems one by one to restore critical functions, and we're trying to use the radio as little as possible so they can work in peace and quiet. We know they'll update us as soon as they can.

Around the middle of the afternoon, Ku-band is restored, and we have the luxury of our phone and our painfully slow internet connection. I ask Yelena if I can borrow her computer to make a quick call to my family and tweet. My close family were in contact with ESA straightaway, and they've had reassuring news, but I don't know what the rest of the world has heard from the media, and I fear my friends and contacts may be worried. 'We're all safe and doing well in the Russian segment,' I write. Hmm. Tweeting in the middle of an emergency means I won't be able to construct a thrilling story of danger and heroic drama about it one day. How short-sighted of me . . . We're still together in the Service Module when it's time for the evening DPC. It'll be another few hours, CAPCOM tells us, because Mission Control is waiting to recover a final batch of telemetry data before they'll allow us to return to the potentially contaminated modules. They assure us, however, that we'll be reunited with our toothbrushes before we go to bed. Houston asks Butch to designate two people to leave the Russian segment and measure the concentration of ammonia in the Lab, while the others wait in their respective Soyuzes, ready to evacuate if necessary. The possibility that we might have to evacuate the Space Station today is remote, we all know that, but given that we're considering this highly unlikely possibility, Butch decides that he and Terry will be the ones to venture forward. That way, each Soyuz will be sure to have an unaffected flight engineer and commander on board. About an hour later, with Houston's authorization, we all put on our respirators and split up. Some of us go up, some down, and some forward. A short time later we finally have the definitive confirmation: there hasn't been an ammonia leak.

When I emerge from the tunnel and leave the Russian segment behind, I have a sudden feeling that something is wrong. Node 1 doesn't smell right; or rather, Node 1 has an odour, and it hasn't had one before. After a few minutes of concerted sniffing I realize what's wrong. What a dunce, I tell myself. Really! The air in here has stood still all day long, and the fan has only recently been restored. What I'm smelling is the clammy odour of stale air: the Space Station smells like an old cellar.

By this time it's late evening, but after a day of waiting and forced inactivity, Terry, Butch and I are ready for action. We collect all the used emergency equipment and put back whatever can be reused; the rest we put in the bin. We spend a long time consulting with Houston about the oxygen masks: how many are left and the best way to redistribute them throughout the Station. We reconfigure some of the power cables so the controllers can restart some of the critical machines. We reboot the odd computer that hasn't coped well with the upsets of the day. At last we find a place to camp for the night. We can't sleep in our crew quarters tonight because Node 2 isn't yet ready to welcome us back: most of the lights and electric plugs aren't working, there's no cooling, and the fans are turned off. The same goes for JEM and Columbus. Butch has put his sleeping bag in Node 3 so he won't disturb anyone tomorrow morning when he gets up early, as usual, to work out. I make myself comfortable in a corner of the Lab. The campsite up here is quickly organized. All we have to do is attach our sleeping bags somewhere, using a simple hook, and the deed is done.

It'll take days to get everything back to normal on the Space Station. Cooling loop B will have to be progressively reinstated with great caution, so as not to risk freezing the water in the heat exchangers, which would cause them to rupture. Then we really would have an ammonia leak!

34.

Dragon has been berthed at the nadir port of Node 2 for a few days, and it's become an extra room of our space home. It's like being able to access a small cellar crammed with shelves – all we have to do is go through a large hatch. The pressurized volume in Dragon doesn't have real shelves, but it is neatly packed to the brim with bags of various standard shapes and dimensions, all securely fixed with thick straps in such a way as to create walls and flooring of a more or less cube-shaped free space; you can move around in it, but no more than that. Even this central volume was completely full when Dragon arrived, and it'll be up to us to replicate the optimal packing and stacking over the next weeks as we slowly reload Dragon with return cargo. This is going to be fun.

The priority cargo removed immediately after Dragon arrived included samples of the Drosophila Melanogaster, or the fruit fly. It seems that they are one of the animal models most used in research, since their genome is well known and they reproduce so rapidly, making them ideally suited for multigenerational experiments. We hope to be able to send back to Earth, conveniently frozen, the children, grandchildren and great-grandchildren of these happy astro-fruit flies, at the moment flying around completely carefree in their cassettes. I assembled their habitat only yesterday. To ensure that we have a control group, only a few flies will be kept in static positions exposed to weightlessness, while the others will be placed in a small centrifuge that will simulate the gravity level of the Earth's surface.

In addition to the fruit flies, Dragon has brought us some actual fruit, a small provision of apples and oranges. We suspect, though we can't be certain, that these are the same apples and oranges assembled for the original launch date a month ago, but never mind! It's a luxury to have fresh food up here. Let's just say we'll eat them quickly. Butch is even happier about his bottle of mustard. For the past month, he hasn't stopped reminding the ground team about his craving for mustard, possibly overdoing it a little on purpose. There's a certain complicity with the ground team over these small things. For my part, I'm happy with the variety of food on board and with my packets of bonus food, particularly the quinoa salad and the bean soup. Like a good Italian, I need only one condiment: olive oil. True, I do miss bread, and the NASA menu offers only tortillas. Happily, there's some real bread in the Russian pantry, and Sasha, Yelena and Anton always put some aside for me. It comes in little cubes so you can put it into your mouth whole without scattering crumbs. They even have my favourite Russian bread, called Borodinsky.

After lunch I go to the JEM for the Zero Robotics final, an international contest for students competing to write software for the Spheres, polyhedral objects the size of a basketball. With small manoeuvring thrusters fed by CO_2, they can rotate and move inside the Space Station. To prepare for the final, we've got a space ready for them inside the JEM, where their position will be detected by special sensors. At the start of any one-on-one run, we position two Spheres in the centre and then let the two teams' code control their manoeuvres. In the game's virtual environment, they have to photograph an asteroid and send the images to Earth, taking refuge as needed during the occasional solar flares. The finalists are following live, some from MIT, some from an ESA centre in Holland, still others from Moscow. Butch, Yelena and I wrestle with amateur commentary, especially when we lose Ku-band coverage and the kids can only hear our voices. This is the beauty of being an astronaut: you end up doing a little of everything, even sports commentary.

This Sunday morning I put my lab technician's hat back on and get busy with the fruit flies. Unfortunately, they're not doing that well. They've deposited their larvae and a few specimens of the second generation have hatched, but they're no longer the happy astro-fruit flies of the first day. The chief suspect at the moment is excess humidity. Prompted by Mission Control, I remove the front panel of the rack hosting their habitat and install a fan. Let's see if this quick fix will save our flying friends. They are not, as it happens, our only multigenerational subjects undergoing experiments right now on the Space Station. Dragon also brought us a culture of Caenorhabditis elegans, C. elegans to their friends. The name denotes a worm about one millimetre long, and this, too, is an ideal model animal whose genome is well known. We'll breed four generations, and specimens of each generation will be chemically preserved and kept in the freezer until Dragon leaves. The experiment investigates the inheritance of epigenetic adaptation, a change that is not linked to modifications in DNA but which influences gene expression, determining which genes are read and actually used in the cells as instruction manuals for protein production.

The morning's work finished, I finally have a bit of time to open my new Crew Care Package, which was sent by the psychological support group. I prepared some of the things in it for myself before leaving, including three books I wanted to have with me in space. Other than the inevitable *Hitchhiker's Guide to the Galaxy*, there's *Mr Palomar* by Italo Calvino and Antoine de Saint-Exupéry's *Flight to Arras*. There's also a nice surprise from Lionel in the packet, originally intended as a Christmas present for Terry, Anton and me. For each of us he's sent a trio of Lego minifigures small enough to hide in your fist, and showing us in our Sokol suits. They are perfect down to the last detail, from patches to name tags – even the hairstyles. Each of us is holding a characteristic object: Terry, with his passion for American sports, has a football; Anton is reading a manual, *Soyuz for*

Dummies, whose cover photo shows him in a restaurant in Cologne with an enormous pork knuckle in front of him; I'm holding my smartphone, a not-so-veiled allusion to something Lionel has always considered an addiction. If nothing else, I've certainly freed myself from that up here, and with no withdrawal symptoms. On the contrary! I don't miss it at all.

Another thing that arrived for me on Dragon was a strange set of nightclothes: a sleep-study vest, which I'm wearing for the second night in a row. It's not very comfortable, to be honest, because it's extremely tight, but that is necessary to allow the integrated sensors to function adequately: electrodes for the classic electrocardiogram and a small 3-axes accelerometer positioned on the sternum for observing the mechanics of the heart, or the opening and closing of its valves. They're exploring the hypothesis that small variations in cardiac function are connected to the poor quality of sleep so many astronauts have complained of over the years. For my part, actually, I feel like I'm sleeping very well up here. It may be down to the fact that during my years of training I bounced back and forth between continents, always jet-lagged, and here I've been living on UTC time for two months. The choice to use UTC time is not due to operational or technical considerations, nor the special appeal of using a universal time reference on a Space Station. Instead, it's said to be dictated by the earthly problem of commuting: it's important that the astronauts' day on board should not finish too late in terms of Moscow time, so that the TsUP specialists can still take the metro home in the evening. In Houston, where there's no metro and everyone drives everywhere, starting a shift at Mission Control at one in the morning is a lot less problematic. We Europeans happily find ourselves in the middle.

International Space Station, 23 January 2015

I'm often the last to go to sleep, the one who closes the shutters for the night in the Cupola, to protect the windows from the impact of small meteorites when human eyes aren't devouring the view from inside. By now I'm familiar with the background noise; it's the normal sound of a healthy Space Station. It was quite different when I first got here. Everything was a discovery then, and I didn't yet know what was normal and what might be cause for alarm. One night, for example, I called Houston to report a much louder noise than usual coming from the floor in the toilet, with vibrations that were felt through the foothold. It seemed like some machine was unbalanced. But it was completely normal: just the noise produced by the urine-processing assembly. Right now it is broken. A few nights ago Houston called me – as usual, I was the only one still up – and asked me to reconfigure the valves so as not to transfer the urine to the recycling assembly. Now it gets collected in an internal container, but we can continue to use the toilet. Last month, when the dosing pump of the chemical treatment fluid broke down, we couldn't. Luckily there's a second toilet in the Service Module, and we all used that for a couple of days. Anton and Yelena never complained, but they couldn't have been very enthusiastic, since they sleep right next to it.

The Russian toilet is actually the original model, and the one in Node 3 is an almost identical copy, just a little more comfortable thanks to the large and well-lit cabin. Yet there is one important difference. The toilet in the Service Module isn't connected to the recycling system. So every now and then I float into the Russian segment for a bizarre example of trading-and-bartering: I appear in the Service Module with an empty container for liquids, and in return I'm handed a full one. Not just any container, but the precise serial number indicated by the TsUP specialists in their 'radiogramme' – that's the name of the documents in which Moscow sends instructions to the crew. In this case, the instructions concern the delivery of precious material: Russian pee for recycling in Node 3. The trading only takes a few seconds, just time for a witty exchange, then I fly away with the container and usually immediately start the transfer into the recycling tank. Like the good commander he is, Butch saw the opportunity to introduce a playful element into this simple operation that would amuse us as well as the controllers on Earth. Per procedure, the urine transfer should be interrupted when the tank is 71 per cent full. Given the slight delay in updating the value on the computer screen, and the need to execute some manual actions in different locations scattered in Node 3, it's difficult to stop at precisely the desired value. Well, for two months now, every time we succeed we gain a point, and Mission Control has been keeping strict accounts. The competition actually finished today. I'm absolutely sure that Butch has won, but the flight director has invoked an obscure rule from some television programme, it seems, *The Price Is Right*, to declare me the winner. The commander can't win – it wouldn't be any fun.

International Space Station, 25 January 2015

Installing a fan wasn't enough to save the fruit flies, and we had to stop the experiment. On the other hand, the worms have grown very well, or so they tell me from Tsukuba. Every time I take them out of the incubator to separate the offspring and put the adults in the freezer, they ask me to show the sack to the video camera. I don't speak to researchers directly, but I know that they are looking on. Up to this point, the colour at least has convinced them that the multigenerational experiment is going well. I really wouldn't know how to tell, because the naked eye can't see the difference between the C. elegans and the rest of the whitish sludge made up of food and small bodies, but I'm definitely taking care of them as conscientiously as I know how. After having lost the fruit flies, I'm even more attentive to my little worms.

When Dragon arrived we finally got the back-up hardware for the Drain Brain experiment, after the original unit was lost last October in the Antares rocket incident. It consists mainly of three bands of stretchy material that I have to wear on my neck, wrist and ankle while doing the experiment. It's a very simple system for measuring blood flow in veins, and one of the objectives is actually to validate this method, which could become a practical alternative to the ultrasound. The second aim is to study the return of the blood from the head to the heart, a mechanism that's influenced by our breathing but also, as you might guess, by gravity. It's another experiment for which I'm the guinea pig: I perform a series of small movements while taking forced breaths, with my nose plugged and my mouth hermetically closed around the mouthpiece of a long tube that's connected to an instrument for measuring lung function. The experiment also requires an ultrasound of the jugular, a fairly simple target.

It's much more difficult to get a picture of the heart's valves at work. Today we were ready to give up, because the time allotted for the experiment was running out, when Butch moved the probe without listening too closely to the instructions over the radio – and the long-awaited image appeared.

'I don't know what you did, but don't move – that's good!' the ultrasound technician was quick to say from Earth. It's not easy to guide someone remotely, that's for sure, especially when it involves images that the technician herself must patiently search for. And then, as far as this particular image is concerned, I'm a notoriously difficult subject. The experiment is called Cardio Ox, and it's looking at a possible correlation between inflammatory and oxidative stress and spaceflight-induced changes to the heart and arteries.

I always get a sense of peace from contemplating the ocean. The Sun is about to set, and the water is molten bronze and gold. Shards of low cloud fly over it, each one pursued by its own black shadow, which is sometimes clearly outlined and at other times fused with the cloud itself, as if a painter had tried to emphasize it by outlining it in black. They are accompanied by flocks of small cumulus clouds, hasty white brush strokes. As the Sun slowly descends towards the horizon, they darken against the backlight, until the golden water is studded with black dots. While the sun is setting, I let my gaze drift over the horizon. Cumulonimbus clouds stand out against the dramatic background of the last warm light of day, heavy with rain and towering over the other clouds. They would be threatening for anyone flying close by, but they're harmless for us, since we're hundreds of kilometres higher than they are. It's the same with cyclones, which we see fairly often when we're flying over the ocean. Sowing death and destruction on Earth, they are a picture of quiet strength from up here. We're sometimes asked to shoot a sequence of photos to aid research activities or the observation of a particularly dangerous cyclone. I have shot lots of photos of them by now, including ones of the so-called super typhoons, which extend so far that they occupy all of the planet that I can see. The other day, however, I managed to take a really unique photo: I captured the eye of Cyclone Bansi while a nearby flash of lightning illuminated it, outlining its contours in the night. A secret, surreptitious shot – and entirely fortuitous.

International Space Station, 28 January 2015

It's almost time. Only a few seconds to go. I let myself go in midair in the Lab, trying not to give myself any push if I can avoid it. Butch and Terry are doing the same thing not far away. Slowly, slowly, we all start to move towards Node 1, as if gently pushed by a mysterious force. In reality, exactly the opposite is happening: while we are suspended and detached from the structure, ATV thrusters are pushing the Space Station and moving it around us. This sometimes happens, and sometimes we don't even notice it if it happens, say, while we're sleeping or concentrating on our work. Usually ATV thrusters are turned on to execute a debris-avoidance manoeuvre, or for a reboost, to raise the Station's orbit. Our orbit inevitably decays over time, chiefly because of the resistance offered by the thin residual atmosphere that still exists even at our height of 400 kilometres. However, today we're doing the opposite: a deboost, or slight lowering of the orbit to adjust the orbital mechanics in view of the upcoming docking of a Progress cargo vehicle. For ATV thrusters to have a braking effect, so to speak, the Space Station has been temporarily rotated by 180° so that ATV is now at the bow. It would have been entertaining to watch the rotation, but unfortunately there are strict rules: during an engine burn, we have to keep the shutters closed to protect the windows from the plume. Since we couldn't look outside, the rotation itself was completely imperceptible.

The deboost over, we go back to our work. Today, I swapped my lab technician's hat for the logistics hat. For several hours now, I've been going around the Space Station with a thick pile of paper, pages of tables filled with tiny characters, some so small that they're difficult to read. This may be because our printer is not very modern; it appears to have come straight out of a 1990s office. These days, Butch actually refuses to call it a printer, since the gadget in question so often refuses to do any actual printing. However, we've managed to extract the long table, a list of all the bags that have to be loaded on Dragon, and details of every item in each of those bags, the current position of each item on the Space Station and any special packing instructions. We call this table the 'cargo message', but since it's too complicated for an astronaut to use (apart, perhaps, from some rare, exceptionally capable specimen), the table is accompanied by further pages of instructions that delineate the order in which the bags must be prepared: the 'choreography message'. The name is not inappropriate, I'd say. Cargo operations are actually a well-choreographed dance carried out over several weeks, at the end of which the vehicle is packed to the last cubic centimetre. Moreover – this is tricky, and also extremely important – the centre of mass will be located precisely where it must be to guarantee correct attitude during the critical phases of re-entry through the atmosphere.

The first version of the choreography message arrived before Christmas and was put on our task list, which planners can use to insert activities that are ready to be carried out, in case we choose to do them in our free time. All of us did so on the weekends and holidays so that when Dragon arrived we had about twenty bags partially packed and temporarily stowed behind a bungee web in the forward endcone of Node 2. It's a good thing that we got ahead of ourselves, because the pace at which we're working is now truly frantic.

International Space Station, 1 February 2015

It's never been easy to work in ATV. For a start, it's some distance away: about 80 metres from Node 2. In these frenetically busy days, when we often don't have the time for a proper lunch break, it's rather irritating to float into ATV only to realize that you've brought the wrong tool or that you've forgotten something you should have taken with you. And then if you have to speak with Mission Control, you have to move into the Service Module, since cargo vehicles don't have radio panels. Right now this means crawling through a half-closed hatch. The problem started a few days ago: in one of the many compartments already filled up with rubbish, a bag has probably inflated due to the gradual decomposition of its contents, and has begun to release gases. We are well aware that we still have some very old bags of moist rubbish on board, months old by now, since Cygnus did not make it to ISS last October. Nothing too dramatic, but the end result of this inconvenience is that there's a rather unpleasant pong inside ATV. Our Russian colleagues, who work and sleep in the adjacent module, have requested that we stop the exchange of air, so we've turned off the fans and are keeping the hatch partially closed.

On the other hand, however difficult it is to work in there, ATV continues to do us a heap of favours. Besides lending us its engines when we need them, it still has air in its tanks, and now and then we open the valves in order to let some air into the cabin of the Space Station. Moreover, when it leaves in a couple of weeks, we'll be free of tonnes of rubbish, including solid waste containers from the toilet dating back to the last crew and huge blocks of packing foam that Butch, Terry and I literally sawed through during one lunch break the other day so they'd fit through the small Russian hatches. It's really great to see the Space Station less crowded with every passing day. Future space missions, further away from the Earth, will surely need to reduce waste material. Additive manufacturing could prove useful, for example, so we could print tools and spare parts as needed, and maybe even recycle the materials when they were no longer used. We're taking the first steps in that direction: the first space 3D printer demonstrator arrived on board a few months ago.

Even used clothes end up in rubbish bags, since there's no way to wash things on board. To optimize their disposal, we often use them as filler material around the larger bags to be loaded on Dragon, neatly packed in Ziploc bags sealed with duct tape. When they're personal clothes, typically t-shirts with some particular meaning for us, we write our name on the duct tape, with a plea to send the Ziploc to Bernadette in Houston. There's no guarantee that the unpacking team will do so – they're definitely under no obligation – but they're rumoured to be generally agreeable and willing to help. That's how I wrapped up a UNICEF t-shirt sent to me on board. I participated up here in a campaign about the needs of children in the poorest countries, doing by far the most embarrassing thing in my life: a horribly off-key performance of 'Imagine' from the Cupola. I was told that they'd only use a couple of seconds of it in a choral performance with many other contributors. But to my horror, I learned that the whole thing is on YouTube.

Oh well. It's all for a good cause.

35.

> And then I fix my eyes on those lights
> that seem pin-pricks,
> yet are so vast in form
> that earth and sea are really a pin-prick
> to them: to whom man,
> and this globe where man is nothing
> are completely unknown
> Giacomo Leopardi, *The Wild Broom*

International Space Station, 10 February 2015

Illuminated by the newly risen Sun, Dragon stands out clearly against the Earth, which is still cloaked in darkness. I'm tempted to say it's hanging off the end of the robotic arm SSRMS, but 'hanging' seems a misleading term to me: it smacks too much of gravity. I could let Dragon go – actually I'll do that in a few moments, and it certainly won't fall to Earth. It will keep floating with us on a nearly identical orbit to ours, only about 10 metres lower.

Here we go: I push the release button, which loosens the firm grip of the end effector. Dragon is now no longer of a piece with the Space Station; it's an independent satellite of the Earth. Soon it will start moving on its own relative to us, obeying the inexorable laws of orbital mechanics, at first imperceptibly, and then more and more noticeably. We won't wait for it to happen, though. I pull the left-hand controller towards me very gently, in order to avoid oscillations, and the robotic arm starts moving, slowly, slowly. When I have visual confirmation that the end effector is clear of the pin, I move the SSRMS away with greater speed, leaving Dragon to float on its own.

There's something unreal about seeing it out there, with its two clearly distinct parts: the unpressurized section, a chunky white can with two extended solar panels, and on top of that, the habitable compartment, bell-shaped and gradually narrowing towards the hatch; we were floating in it until only yesterday. There are almost two tonnes of cargo inside, all neatly packed, the result of a month's intense activity here on the Space Station and thousands of hours' work done by many people on Earth, most of whom we don't know. Who knows how many eyes are trained on Dragon right now, watching its departure with a sense of involvement that goes far beyond simple curiosity? And how many will be following its re-entry through the atmosphere, waiting anxiously for news of its splashdown? For our part, we'd be very interested to hear what the specialists think of our work. We hope it will be something like, 'Wow, look how well Dragon is packed! Not a single loose strap, not a bag out of place.'

'Indeed! And the work Terry and Butch did on the hatch with tape? Exceptional! Not a drop of water got in after splashdown.'

Well, you'd have to be a fly on the wall in the meeting room, since if there are any loose straps or untidy bags, they might not tell us. I get the sense that the people communicating with us from Earth are quick to compliment us but reluctant to point out our mistakes unless they have some immediate impact on the mission. It's not that they're obsequious – definitely not – but I think they try hard not to make us feel guilty or embarrassed. They don't want us to worry. I understand this, and it's no doubt a good thing, but sometimes I feel like they say, 'Great work!' too often. In any case, eventually we'll be getting honest feedback, ultimately from the astronaut evaluation board, after the mission.

'Go for Dragon depart,' I say to Terry. I've retracted the robotic arm, and it's now at a secure distance of 4.5 metres, to judge by sight. Terry sends the departure command, and Dragon turns on its thrusters for its first departure burn. It moves slowly against the background of the western Pacific, over which the Sun is unfurling its light. Only yesterday it was still part of the Space Station, and now it's an inaccessible shell, a delivery from space to

Earth, carefully prepared by the only human beings not currently located on the planet.

I must say, I actually feel some relief watching Dragon leave for Earth.

The past month has been a sort of mission within the mission, with long working hours. I'm always quite tired at night here on the Space Station, maybe because we have such regimented days, and maybe also because of the elevated CO_2 levels. But lately I'm exhausted by the end of the day. For Butch and me, the final surprise came on Sunday morning when we removed an FPS (Fan Pump Separator), a critical component providing circulation of air and cooling water in the EMU. That FPS had failed during a routine test of the suit the week before, and we had to send it to Earth urgently on Dragon so the issue could be investigated.

Butch and I aren't new to this work: last December we replaced the FPS in the suit Terry will use in a few weeks. It's a delicate matter: for reasons that are not entirely clear, these units have been breaking down frequently. The idea of replacing them in orbit would probably have made EVA specialists faint from shock a few years ago, when all suits were serviced on Earth by highly trained technicians working in a clean room. When the Space Shuttle was retired, the possibility of regularly rotating the suits disappeared with it. They had to try and make the best of things and let the astronauts on board carry out periodic suit maintenance as well as less critical repairs. Until, a little while ago, someone replaced an FPS for the first time, a procedure often described as brain surgery on the suit. On Earth, it would probably be a maintenance operation like any other, somewhat long and complex, but manageable. Up here, though, we have to grapple with weightlessness. While most of the equipment we use on board was conceived from the outset for maintenance in orbit – for example, using captive fasteners that unscrew but don't come loose – the EMU suit was heartlessly designed to be disassembled only where nothing floats. Packed as it is with tiny, almost inaccessible screws and a half dozen washers each, losing one – which could float into the suit and remain hidden inside it – is all it would take to put your spacewalking colleague in danger. So it hardly needs saying that we take all necessary precautions: we use sheets of lint-free paper to cover access to the inside of the suit; we aim a vacuum cleaner at the site we're working on and cover the hose with a piece of fabric so that any stray screws or washers floating around will be caught; and we transmit live images of our work to Houston, where a room full of specialists sit and watch attentively. Replacing the FPS in Terry's suit last December took a day and a half; we needed several hours just to gather all the specific tools. To my great relief, a series of checks the next day confirmed that it was working well. The suit is ready for Terry's spacewalk in a few weeks. It's going to be another hellish time, with three EVAs in fewer than ten days. A heavy burden for Terry and Butch, certainly, but I know very well that they're going to be difficult days for me, too, and not only professionally. More than anything else, they'll be difficult emotionally.

International Space Station, 15 February 2015

As expected, Dragon splashed down off the coast of Los Angeles a few hours after leaving the Space Station. Thanks to one of Terry's friends, we received photos, almost in real time, of the capsule suspended from three parachutes at the moment it splashed down in a rather turbulent sea. Seeing the photos gave me a sense of 'mission accomplished', at least for now. The next Dragon will arrive in a couple of months, and this time I'll be at the command of the robotic arm to capture it.

With any luck, we'll be able to take photos of the end of ATV-5 Georges Lemaître's mission ourselves. It undocked yesterday after a few suspenseful days due to a failure in one of its four power distribution units. As a European astronaut, it was down to me to close the hatch of this last ATV, symbolically putting an end to the programme, but Butch is the uncontested hero of ATV-5. He took it upon himself to dispose of the greatest possible amount of the rubbish and packing material, and he often worked tirelessly in his free time filling up any re-maining space, always coordinating with ATV Control Centre in Toulouse. The vast internal volume of ATV-5, so empty a few months ago that, as a newcomer to space, I struggled to find enough handholds, was so full by the time of its departure that the six of us could barely get into it for a souvenir photo.

Butch is in the Cupola with Terry and me, and we're all intent-ly focused on the horizon. Georges Lemaître is re-entering the atmosphere at this very moment, and Houston has told us that if we look in the right direction, we should be able to see the smoke trail the instant it begins to disintegrate and burn up. Only a few minutes ago, Toulouse sent the last command to ATV, to turn on the thrusters that will make it spin, in such a way as to offer maximum resistance to the atmosphere. ATV's last act is a con-trolled suicide. Rather than a smoke trail, what we see are two white puffs outlined against the black of space and far above the variegated strokes of blue and azure that envelop the planet. The atmosphere, we know, is much thicker than that thin, luminous mantle. It's in the upper atmosphere, invisible to the eyes, that Georges Lemaître has met its fate.

The departures of Dragon and ATV in rapid succession have made me think about how much of my time on board has already gone by. I realize, with some anguish, that I've almost reached the turning point of my mission. In three months – barely three – I'll be back on Earth, dragging my weight around effortfully on my legs. And the Space Station will continue its laps, a dot in the evening sky, heedless to the absence of the umpteenth earthling who stopped here and stayed for a while, no different from those who came before or those who will come after her. Other astro-nauts will take care of humanity's outpost in space.

The Space Station isn't on a journey. There's no departure and no arrival. The on-board activities go beyond the limits of a sin-gle expedition, and every crew takes the baton from the previous one and passes it on to the next, in a relay that has been going on now for nearly fifteen years. Fifteen years of uninterrupted human presence in space. Or at least fifteen years in Earth orbit, because when it comes down to it, we're all in space. The space that bends and expands, suffused with a melody reaching us from about 14 billion years ago in the form of cosmic background radiation. How our vision of the universe has changed in a few hundred years, or just in the last century! This very year, we're celebrating the hundredth anniversary of the theory of relativity, which radically changed our understanding of space, time and gravity. I'll never cease to be amazed at the capacity for imagination or abstraction in the most brilliant human minds. The rest of us try to plod along, slowly. Already in ancient Greece, the astronomer Aristarchus developed a heliocentric model, or the idea of the solar system with the Earth rotating around the Sun. And Eratosthenes was able to calculate the circumference of the Earth with surprising accuracy. More than 2,000 years have passed and a few – though happily, very few – still believe the Earth is flat. That's what we humans are like: some of us are capable of scaling the heights of rational thought, and others are imprisoned by delusions and fallacies. Most of us fall somewhere in between, sometimes winning, other times losing in our daily struggle to think clearly.

In all this, the scientific method is a formidable ally, with field-tested value. We can only admire the predictive capacity of contemporary science, which has provided the foundation for fast and furious technological development, to which we owe the comfort of our homes and an average lifespan that has nearly doubled in just a century. If we can manage to placate the demons of the human heart, to survive our own power to create and destroy, what else might we come to understand about the fabric of the universe? Will our children and their children smile at our actual convictions – as we do at our parents' naivety in these matters? Will future generations be any better at reconciling humanism with our marginality in an indifferent universe? With our living in a solar system like many others, on one peripheral arm of a galaxy like so many others? When I looked at the sky as a child, I was simultaneously frightened and fascinated by the staggering number of stars in the universe. Thousands of billions of billions . . . how can we grasp the meaning of such a number? These days I know how to write it as a power of ten, how to think of it as order of magnitude, and how to represent it on a logarithmic scale. But how can I truly grasp it, even when I strain my small human brain? Is it possible to conceive 14 billion years, the estimated age of the universe, more or less, when the span of my own life is at most a century, and my species has handed down its own history for only 5,000 years? We're just the blink of a star. They die giving birth to the atoms we're made of, but they know nothing of human doings and are indifferent to our flashes of greatness, and the depth of our egos. Perhaps if we looked at things from a cosmic perspective, we'd be more likely to forgive each other our pettiness, to help each other live peacefully during our brief time on Earth.

36.

No visiting angel, or explorer from another planet, could
have guessed that this bland orb teemed with vermin, with
world-mastering, self-torturing, incipiently angelic beasts.
  Olaf Stapledon, *Star Maker*

International Space Station, 11 March 2015

We say goodbye to Butch, Sasha and Yelena in front of the open
hatch of their Soyuz. We're in the MIM-2, one of the small, spherical Russian modules, and so are they, but they're packed into the
cramped, square transitional compartment, and with their legs
pointing towards the hatch of their Soyuz, they're already nearly gone. They're dressed as usual, in t-shirts and cotton socks,
nothing that would make you suspect that they're about to pass
through the atmosphere in a ball of fire. Even though it's all part
of our work, and we like to think of it as routine, it's not actually
the safest way imaginable to go back home after a business trip. I
give them a warm hug, feeling a twinge of apprehension. When
they've said their goodbyes, they float backwards to enter the
Soyuz, as if letting themselves being sucked back into the womb.
  Some hours later, I go to listen to their undocking through
the closed hatch. I can't see them leave, anyway. My only chance
would be to watch them through the little window of the Node
3 aft hatch, but Terry has set up his camera there with the interval timer, as we often do with photos of Earth. That way, Butch,
Yelena and Sasha will have a time-lapse video of their undocking.
That's it. A few dull thuds come through the metal that separates
us – and they're gone. On the other side there's nothing now but
the vacuum of space.

I join Terry in Node 2, where he's prepared the bell. As a good US Navy officer, Butch asked us to ring it according to Navy ritual, to signal the departure of the Soyuz. We transmit its ringing over the radio, along with our best wishes for their re-entry. Barring another mission together, which is unlikely, from this moment on, our lives, so closely interwoven over the past few years, will diverge for ever. Of course, we won't lose sight of each other completely. Astronauts always remain connected in some ways, as if by elastic threads that periodically draw some of us to the same place for a launch, a commemorative event or a get-together. So we're bound to see one another even if, as I expect, some of us decide it's time to change jobs.

Anton, Terry and I go our separate ways to take care of our tasks in the Space Station, which feels suddenly bigger, but we continue to listen to communications between the Soyuz crew and TsUP coming over the loudspeakers. We spontaneously meet up again in Node 1, after the voices of Sasha and Yelena reporting on the engine's operating parameters during the braking burn first become garbled and finally fade away completely. The silence is premature: contact is usually lost a bit later, during the descent through the atmosphere, when the Soyuz is enveloped in plasma. A call to Moscow is certainly justified in order to get updates. Yet not even TsUP is receiving from the Soyuz crew. Whatever is happening at this moment on board Soyuz TMA-14M, a projectile headed towards the Kazakhstan steppe, is known only to our friends Sasha, Yelena and Butch. We're not alarmed, but we're definitely anxious to have confirmation that all went well. When we start thinking it's time to hear something or we might start to worry, Moscow finally tells us that the search and rescue teams have made contact.

A little later, we see the scorched hull of the descent module, which landed upright, on NASA TV, via a computer conveniently set up in Node 1. As long as there's Ku-band coverage, Houston can send us live TV. Except on rare occasions, we have until now always requested an American sports channel, which Butch watches voraciously. I'm sure this won't be changing: I've heard that baseball season is about to begin, and Terry is a huge fan. I don't have any particular favourites and I don't really like television, even on Earth. After a few tries to get international news, which turned out to be too complicated, I've good-naturedly accepted the background of sports news, with the unrelenting stream of interviews with trainers and athletes, as well as a thrilling in-depth, behind-the-scenes story of a deflated, partly deflated or in any case under-inflated football. Look: if I hadn't ventured into space, I wouldn't have known any of this.

Coincidentally, Expedition 42 actually ended on Douglas Adams' birthday. Yesterday Butch officially handed over command of the Space Station to Terry, and then together we put our mission stickers in Node 1, next to those of all the earlier expeditions, and in the Airlock: a permanent reminder that our crew carried out some spacewalks. Butch and Terry did, to be exact, but if we are talking about the Airlock, I believe I provided an honest contribution. Being in charge of the operations before and after the EVAs was the most challenging duty I performed on board. When we started at 6.45 in the morning of the first day, I took over an Airlock brimming with complex equipment; two EMU suits, notoriously capricious; and two astronauts anxious to go out as soon as possible. In addition, I pictured a little army of EVA specialists on the ground, their eyes glued to the images transmitted from the Space Station, and they in turn wanting me to carry out the operations as quickly as possible – but without making any errors, please. Any mistake would cause a delay in the best-case scenario – or in the worst, spell danger for Terry and Butch. I had dozens of pages of procedures in my hands, printed and annotated, studied and revised again on the night before the EVA, which was an unexpected free day since specialists were still analysing the defective FPS sent back to Earth on Dragon to see if there might be a similar risk of malfunction with the other suits. They concluded, within reasonable engineering certainty, that if the FPS started up normally it would not shut down.

According to procedure, suiting up and pre-breathe take five hours, but this doesn't include any margin, so we decided to start earlier than scheduled. Still, that first time, Terry and Butch went out at 12.45, half an hour late, due to my inexperience – I was double-checking every step – and to some difficulties on Earth in receiving telemetry from the suits. So I didn't manage to get them started on time, but at least I didn't make any mistakes, and they all agreed, I think, that that was the most important goal. Terry and Butch began their subsequent spacewalks ahead of time.

It wasn't easy for me to watch them going out into open space three different times. After five or six hours of absolute concentration, with my hands moving over the suits and equipment in the Airlock like those of a concert pianist on the keys of the piano, I found myself looking through the window of the closed hatch at the cramped, depressurized compartment, a cylinder normally used as a storeroom and crammed with stuff, but now completely empty and exposed to the vacuum of space, abandoned like a room whose occupants have left hurriedly, taking everything with them but forgetting to close the door. Each time, the relief and satisfaction of having carried out an important responsibility was mixed with the bitterness of frustrated desire, a few twinges of regret that were enough to cloud my mood, despite the fact that there was a lot to be happy about with that series of spacewalks: they went without a hitch. Yet I had spent too much time training underwater to do just what Terry and Butch were doing, not to imagine myself out there. They would not have minded swapping places with me. Of course, they wouldn't have wanted to trade on their first spacewalk, or the second, but I don't think either of them was all that keen to make a third or, in Butch's case, a fourth. And yet, as everyone knew, it was impossible. Even if there had been other EVAs, the scene would have been the same: at some point, I would have found myself behind a closed hatch, tidying up the pressurized portion of the Airlock, and waiting anywhere from six to eight hours for their return, during which time the occasional clomp of their boots or clink of their tools on the hull would have reminded me that someone was out there.

With only half its crew, the Space Station seems even bigger. Every now and again, I'm surprised to find myself glancing towards the Service Module, my gaze aligning with the sequence of open hatches, looking for some movement.

Now that there are only three of us, Anton works on his own all day in the Russian segment, where there aren't any cameras transmitting images to the control centres. If anything were to happen to him, we might not realize it for some time. It's just a little additional attention; I'm not particularly worried. Given the focus on safety, an accident on the Space Station is very unlikely. We can't, of course, completely exclude medical emergencies, and we have the equipment we'd need to intervene just in case there were, for example, a cardiac arrest. We're all trained to perform CPR, to administer life-saving drugs with an intraosseous device and to use the defibrillator and respiratory support equipment. A bench is firmly fixed to the deck of the Lab, right in the middle of the cabin, and it's furnished with a number of straps for restraining someone while you're administering CPR. Fortunately, it's never been needed. It's anyway quite difficult to do proper chest compressions in orbit. You can't use your own weight to provide good pressure as you're taught to do on Earth, and you have to find a way to stabilize your own position, which is why the CPR bench is equipped with straps for the first-aider as well. But, like many, I opted for another technique. If I had to administer CPR to a colleague, I'd position myself upside-down above them, my arms raised over my head and pressing against their chest, while I pushed against the ceiling with my legs. It's one of those things you can't try on Earth during one of the Megacode simulations, but it works wonderfully, at least on the bag of towels we used during our on-board practice session. We periodically review the other procedures in the medical manual as well, such as suturing a wound or treating a collapsed lung, but in that case we review only the sequence of the operations, without handling the equipment. There's no question, in fact, of any immediate action. On the contrary: in such cases we wouldn't do anything without the assistance of a doctor on Space-to-Ground. To this point, with all we've learned during our crew medical officer training, we've had to put only one simple, painless procedure into practice: every month we give each other a quick check-up, taking a picture of the eardrum and measuring simple vital signs such as temperature and blood pressure. Fortunately, I've always been well. My only issue has been an annoying muscle ache a little while ago that stopped me doing my scheduled workouts on ARED for a week.

Over the past few days we've been getting ready to welcome Scott, Misha and Gennadi, who are now in quarantine in Baikonur. Just yesterday Terry and I searched the inventory to find out where the sacks containing Scott's clothing were being kept, and we put them in one of the three large semi-rigid bags secured in the overhead vestibule of Node 2; it's the equivalent of a wardrobe or a chest of drawers, and there's one for each of us. That's where I go once every two weeks to get the so-called 'brick', a Ziploc containing two cotton t-shirts, three pairs of socks, seven pairs of knickers, a camisole and two training outfits. Our Russian colleagues can change their clothes more often so there's some good-natured teasing, but our wardrobe allowance seems ample to me.

Almost ten days have passed since the departure of the Soyuz, and I'm just about getting used to not meeting Sasha in mid-air as he comes and goes from his crew quarters in Node 2, or not passing the time of day with Yelena when she's training on the ARED, not hearing Butch talking on the radio, with his contagious cheerfulness and typical southern expressions. 'Bless my heart,' he was always saying, when he realized he'd slipped up on something. I'm wondering how he's adjusting to being on Earth again. Fitting back into family life can't be a simple thing, taking one's place in a family routine and a web of relationships that have been woven around one's absence. There are still pressing demands at work and business trips to make in the six months after returning, and even the most devoted family can struggle to provide that unconditional support at this point, after an astronaut has realized the dream of a lifetime, often at enormous cost and sacrifice to their loved ones. Of course, they understand the importance of the post-flight experiment sessions and the weeks of detailed debriefing on every aspect of the mission, that it's essential to meet with the public and representatives from various institutions, and to take part in popular television programmes. Yes, it's important to talk about it all and get people involved, explain and generate excitement, do what you can in exchange for the exceptional opportunity you enjoyed, 'but please, do the washing up and take the rubbish out first!'

Butch wrote to us the other day to recommend that we do lots of stretching to prevent sore muscles after our re-entry. Stretching isn't part of our prescribed daily training routine, but it seems to me that it's no less important than running or squats. It may not maintain bone density, but it can help to preserve the balance of our muscle chains. I have a strong sense that my hamstrings and my quads are shortening, probably due to the neutral position the body assumes in weightlessness, with the legs a little bent, as if they were dangling over the edge of a wall. Fortunately, stretching in weightlessness is very easy. Lifting up your straightened leg in front of you, for example, until you feel the tension in your muscles requires zero effort. I've made a habit of doing that after every lunch break, when I watch the previous day's edition of the news programme *Tagesschau* if my American colleagues aren't around. The ESA crew support team still sends it from Monday to Friday as they did for Alex, and in the absence of any other requests from me.

The perspective is naturally very European and more specifically German, and it doesn't escape me that being interested in a small corner of the Earth while I'm up here in orbit is a bit of a paradox. It's true that I come from Europe and I haven't left it for ever. Most of my loved ones are there, my work is down there, my country is there, and so is the country where I've studied and now live. It seems to me completely natural that I should maintain an interest in matters in that part of the world. At the same time, everything that makes those matters different from those of the other billions of humans – all that is a difference of culture, history, habit, material conditions . . . all of it seems like a detail to me, an afterthought, a final, hurried layer of paint that adds or removes little of the underlying substance, that kernel of human nature made up of feeling hot and cold, being hungry and thirsty, experiencing joy, pain, fear, wonder and ecstasy. 'Who knows what unusual emotions you must be experiencing up there,' many have written during my months in space. But the palette of human emotion has always been the same. I haven't discovered any new emotions by being in space. It's true that I have experienced moments of particularly intense feeling, but certainly not a different level of consciousness or awareness, as some may imagine. I don't think I've been through anything that distinguishes me from other human beings, other than having an exclusive experience. Instead, I have a sense that each day brings me closer to feeling that I have something in common with men and women of the past, present and the future, closer to recognizing a little of myself in everything that's human, from the peaks of generosity, dignity and courage to the troughs of cruelty, selfishness and cowardice. It could be because the entire Earth has become my backyard, and by this time I recognize the continents even before looking out of the Cupola, simply by noticing subtle differences in the tone of the light that shines through the windows. Maybe because the ups and downs of daily life on Earth seem more and more illusory, and important only due to some perception bias, like small whirlpools, which dissipate energy and change the destiny of the small creatures engulfed by them but don't affect the deeper currents of water. Maybe because up here, the fates of not only single human beings, but of humanity itself, seem precarious and transitory. Earth seems eternal when I see on its surface the scars of processes that have lasted for hundreds of millions of years: impact craters and volcanic craters, lines of collision and lines of separation, signs of erosion and deposition, ongoing changes too slow to perceive. Against this background, everything humans have produced, from pyramids to skyscrapers, from cave-paintings to the works of Picasso – all of it seems to have been squeezed into a momentary pause between two gusts of wind.

And yet I like seeking them out, these traces of the brief story of humanity, the material signs of our creative genius. Before the mission, I resolved to take pictures of as many UNESCO world heritage sites as possible, and NASA's Earth observation group kindly put them all in WorldMap, software that tells you the hour you'll fly over a certain location in advance. Some of my ESA colleagues have also taken my project to heart, and every morning they send me lists of all the passes for the day.

Over the past few days, many people have also sent me directions and suggestions for watching the solar eclipse that affected the whole of Europe this morning, at least partially. We don't have the right filters on board for protecting our eyes, and our orbit only allowed us to see a partial eclipse for a few minutes, but it was intriguing to catch sight, in the distance, of the shadow of the Moon, or its umbra, projected over the Atlantic Ocean. Since its orbit is a thousand times farther than ours from the surface of the planet, the Moon doesn't look very different, up here. Its reflection on the Earth, however, frequently creates captivating tricks of light: one minute there's nothing and then, suddenly, silver sparks flicker over the dark cloak of night, and for a few seconds, fragments of frosted glass betray the presence of bodies of water, soon swallowed up again by the world's black stillness.

37

But nothing the peoples of Europe have produced is worth the first known poem that appeared among them. Perhaps they will yet rediscover the epic genius, when they learn that there is no refuge from fate, learn not to admire force, not to hate the enemy, nor to scorn the unfortunate. How soon this will happen is another question.

Simone Weil, *The Iliad or The Poem of Force*

International Space Station, 17 April 2015

'Dragon has started its approach from 30 metres. Monitor according to step 5 of procedure 1.102.'

Our astronaut colleague David, with his slight Quebec accent, confirms from Houston what Terry and I have already seen from the robotic workstation in the Cupola: Dragon is starting its final approach. Within about twenty minutes it will arrive at the capture point, 10 metres from the Space Station and only 5 metres from the free end effector of the SSRMS arm, which waits, still folded, like a half-closed compass. Soon, it will extend under my commands to reach for its prey: a tame beast, or so I hope.

Before holding at 30 metres, Dragon stopped at 250, and before that, at 350. During this slow process, we must simply confirm that it remains inside the approach corridor and that the correct lights are illuminated on the remote control panel. We diligently report our observations to Houston, as we learned to do in training in a simulated Cupola, inside a large room at JSC that was kept dark, with an image of Dragon projected on a black background that looked like space. In truth, I suspect that if there were some anomaly, they'd become aware of it much earlier in

Houston, or in Hawthorne, California, at the SpaceX control centre. But Terry continues to make regular reports in his role as M2, the support function I carried out for Butch during last January's capture.

Dragon is so stable that it seems completely still; you barely notice the short burns of the attitude thrusters. Little by little, as it approaches, it's easier to make out the details: the grapple pin, the strobe light, the thrusters' exhaust nozzles. This precision encounter between two vessels on a boundless sea never fails to move me. No matter how big the Space Station may be, humanity's outpost in space is nothing but a grain of metal in the vastness of low Earth orbit, and yet Dragon has found us, and is about to start flying in formation with us. Even though it's all planned down to the last detail, this vehicle joining us from Earth with its slow, steady pace, this messenger and bearer of gifts radiates a poetic, even epic aura. We watched its departure live on NASA TV three days ago, and now we're welcoming it at the other end of its voyage. We are the first humans to set eyes on it since it disappeared on top of the Falcon 9 rocket.

I'm trying not to get too absorbed in my thoughts, or seduced by the sight of Dragon. I'll need to stay focused. I'm acutely aware that being in command of the SSRMS today is a chance to become famous – extremely famous – for all the wrong reasons. A slip at the controls at a critical moment is all it would take to induce an uncontrolled oscillation; and pausing just a moment too long before moving the arm away to let the oscillation dampen out could damage the grapple fixture. There isn't another one, so the entire mission, with its two tonnes of cargo, would be lost. This sort of situation must be what our first boss had in mind, a veteran astronaut who never tired of reminding us Shenanigans when we were new recruits: 'Don't become famous!' So today, all I want is for the capture to remain an anonymous event, not worthy of even a footnote in the history of the Space Station.

I can't say I'm worried, in any event. I've performed hundreds of captures at the simulator, and dozens just in the past few weeks, on board, thanks to the Robot software, which is installed on one of the computers in the Lab. Every time I went by, if I had the time, I'd stop for a few minutes to perform a capture, and then I'd load a new scenario while I floated away towards my duties.

Dragon has concluded its approach, and it's now flying in formation with us. Terry and I are ready for capture, but David lets us know that we still need to wait five or six minutes before proceeding, and I imagine one last round of Go-No Go between the specialists in Houston and Hawthorne. While we await authorization for capture, we enter orbital night. Dragon now stands out against the darkness, illuminated by the external lights of the Space Station. The internal lighting of the Cupola is turned off to prevent reflections, and our faces are lit only by the spectral glow of our computer screens.

'You are Go for capture sequence,' David informs us. It's time. With Terry's help – he reads the procedure, though we both know it by heart – I configure the arm in manual mode and set the end effector for capture. On Dragon's remote control panel, Terry arms the command 'free drift'. He'll send it when I've brought the robotic arm to within a couple of metres, and from that moment we'll be working against the clock, because Dragon is no longer actively controlled and might slip out of the capture envelope.

I make some small, on-the-spot rotational adjustments with the right-hand controller, so as to start with optimal alignment, and then I push the left-hand controller forward, initiating the approach. Terry calls out the distances. He's judging them by sight, based on the orthogonal images from the video cameras. He is responsible for monitoring the whole of the robotic arm, because my eyes are glued to the central display. The image comes from the video camera installed on the end effector of the SSRMS, and on it I can see the target, which guides me towards the grapple pin, along with a series of superimposed alignment cues. I have to make only minimal corrections; Dragon is very stable.

'Two and a half metres,' Terry says when we're about halfway through.

'Ready for free drift,' I respond. 'Copy: I'm sending the free-drift command.'

When lights on the remote control panel confirm reception by the vehicle, Terry announces, 'Dragon is in free drift.' I'm therefore authorized to continue, with no need to slow down. I'm glad the timing has worked out: it's better to maintain a constant speed of approach, since every time you push on the controls there's a risk of oscillation in the arm. It's all perfectly aligned.

'One pin,' Terry says, referring to the remaining distance.

'I'm Go for capture,' I tell him, responding to his implied question. We're nearly there. The visual references tell me that we're in the envelope, but I force myself to wait a few seconds. Three, two, one . . . 'Capture!' I exclaim, pressing the trigger under my right index finger, and immediately releasing the controls.

At this point, the robotic arm is completing the capture sequence automatically. Terry and I watch the telemetry closely and keep the malfunction procedures handy, just in case. 'Good tension,' I say at last, reading a value that confirms that the capture device on the robotic arm has a solid grip on Dragon. We won't become famous, at least not today.

'Dragon capture complete. You are Go for post-capture reconfiguration,' Terry reports to Houston while I switch the robotic arm to safe mode. Houston will take control of it later for berthing. After high-fiving with Terry, I grab the microphone to congratulate the ground team for a vehicle and a mission that have been impeccable so far. Then I tell Terry that I need his help for a special photo op and I ask him to wait a few minutes, while I go and change. On board Dragon there's a special, long-awaited delivery, and I have just the right outfit for the occasion.

When I decided to bring a uniform from *Star Trek: Voyager* with me on the Space Station, I was hoping that there would be some way to celebrate the TV series' twentieth anniversary in space. In 1995, I was in the United States as an exchange student, and as a Trekkie since childhood – in a pre-internet era and in a country more or less uninterested in the series, where its episodes arrived late or never – I was delighted by the possibility of watching the saga's new incarnation from its very beginning on TV. As an ambitious seventeen-year-old with an interest in science, I certainly noticed the fact that both the *Voyager* command bridge and its machine room were led by women, a scientist-captain and her rebellious and brilliant chief engineer. In honour of Captain Janeway's passion for coffee, I followed a Trekkie friend's suggestion and got Terry to take a photo of me in my *Voyager* uniform against a background showing Dragon in the firm grip of the robotic arm; and in its belly, the very first space espresso machine.

I can't say to what extent my passion for *Star Trek* really influenced my choices, and it may seem excessive, even childish to claim that a TV series made a significant difference to my life, but sometimes a reaction needs a catalyst, and in my case, the adventures of Mr Spock and Captain Kirk may well have acted as the catalyst for my dreams of going into space. A dream is concentrated imagination, with the power to feed your daily motivation, but also your stubborn determination to pursue unlikely, long-term objectives. Like all that is human, it has its risks. Pride can lead those who achieve their dreams to delude themselves into thinking that success was just the result of their actions, forgetting the lucky breaks and all the chance involved in life – and that life is often less favourable to people who are every bit as deserving. Someone looking in from the outside might be convinced that the lucky person had special talents and a blueprint for success; just look at the countless books and interviews in which people considered successful are asked for their advice. But what about all those of equal talent, and who've worked just as hard without the same success or rewards? They remain invisible. I don't have to look far to spot fate's arbitrary nature. The 2008 astronaut selection would almost certainly have been held three years earlier if not for the Space Shuttle Columbia's fatal accident, which delayed the launch of the European Columbus module. Three years earlier, and I wouldn't even have been able to participate.

I feel it would be a good thing for us to start recognizing again the role fortune plays in human events; to regard the successful with tempered admiration; and with equal admiration those who pursue their dreams with dignity and effort. Even when circumstances aren't favourable, their effort is noteworthy and bears fruit. People who are motivated to give their best every day and to choose the road less travelled in the knowledge that it offers greater opportunities for growth are more likely to enjoy a fulfilling life, even if the dream that spurred them on is never realized. I wish every child in the world could grow up without knowing danger, violence, trauma or poverty. And then, I wish that all of them could have a dream to cherish.

Towards mid-afternoon, once Dragon has docked and the bolts securing it to the Space Station are in place, Houston gives me the green light to leak check the vestibule, the sealed volume now existing between the hatch of the Station and that of Dragon. When I've confirmed that there aren't any leaks, I go on to equalize the pressure and open the hatch Station-side. As soon as I get the final barrier between us and Dragon sliding on its tracks, I'm struck by a pungent, unmistakable odour, coming from the material that's been exposed to open space. The odour is not at all pleasing – a slight staleness, with undertones of something burned, like the smell you get after welding. Somewhat inappropriately, we call it 'the smell of space'.

I'll spend a couple of hours today reconfiguring the vestibule and cautiously, meticulously removing a number of electronic units for controlling the berthing mechanism. We'll open Dragon's hatch tomorrow morning and immediately retrieve the urgent cargo. Top of the list for me is a small Kubik incubator I've got to install in Columbus so I can begin the Cytospace and NATO experiments. Cytospace looks at the effects of weightlessness on the cytoskeleton, the structure that gives form and mechanical support to the cells. NATO will test the hypothesis that adding certain types of nanoparticles to bone tissue can effectively counteract osteoporosis. The cell cultures will stay in the incubator for a few days before I put them in one of the freezers. After a month, Scott will take care of transferring them to a cooler bag for re-entry on Earth. Terry and I will have been gone for several days by then.

I can hardly believe it, but there are only four weeks to go before our re-entry. I have the disturbing feeling that time is running out whenever an activity connected to our return is put on the timeline, as it is more and more often, whether it's preparing our personal data for download or checking that the seats in our Soyuz haven't become too small for us due to the lengthening of our spines. I'm not looking forward to returning. On the contrary. Of course I'll be happy to hug the people I love, to eat a big fresh salad with cherry tomatoes and mozzarella, to take a real shower, and feel the water running through my hair . . . But I'd be willing to delay my return and all these pleasures for a little bit more time in space. I like being up here. The excitement of the first few weeks has settled into a quiet fondness. I feel comfortable with my routines on board and fulfilled by my activities up here. Even when they seem trivial, they bring to fruition the concentrated efforts of many people. Then too, there are the small pleasures of living in space, from somersaults in mid-air to breathtaking views from the Cupola. Even after almost five months on board, there's always something new. For example, I haven't yet seen the noctilucent clouds, wispy clouds that form high in the atmosphere and whose delicate glow becomes visible only after sunset, when the Sun illuminates them from below. So I'd like to stay, at least until I've seen the noctilucent clouds. It would be a crime to leave before that.

I've been joking for weeks with Anton: do the Russians really want us to come back on 13 May, so close to Victory Day celebrations? Hasn't he heard something from his TsUP acquaintances or Roscosmos about a possible reschedule? Obviously, these are light-hearted exchanges, quips for passing the time. There's actually no reason to expect a delay. In less than four weeks, we'll go back to life on Earth, with all its demands and complications. We'll have to get used to leaving the house every morning, preparing a backpack with our stuff for the day, doing the shopping and deciding for ourselves how to organize our time – at least in the gaps left by our packed post-flight schedule. If one of our kitchen appliances breaks, we won't be able to call Houston to tell us where to find the spare parts and the repair procedure. Let's be clear: apart from the strictly technical aspects, and we're very well trained for those, life up here is pretty simple. Full of constraints, but also linear, pure and crystal clear, so much so that you can lose touch with all the ambiguities and irreconcilable contradictions that are a part of life on Earth. For example, does society owe something to imprisoned young offenders, or do they have a debt to society? Are they victims or perpetrators?

I'll be speaking to a group of them tomorrow, thanks to the ham radio on board. Some of the astronauts are ham radio enthusiasts and they use it often and spontaneously, to the delight of the earthlings who yearn for an exchange with the Space Station. I've only made myself available for school contacts, which are made possible by an international association of tireless volunteers. Besides overseeing the organizational and technical aspects, they prepare the pupils in a series of preliminary meetings, collecting their questions and sending them to us beforehand. As it happens, the audio quality isn't always high, and it's a shame to make someone repeat a question, losing precious seconds of what little time we have – about ten minutes, when the Space Station is visible from the ground station in use.

The kids I'll be speaking to attend school at a young offender institution and, judging from the questions they sent, it seems they have found many similarities between life on the Space Station and life in prison. I know they've probably inflicted suffering on others, and yet I can't help but feel some tenderness towards them, a strong desire to reassure them, tell them that they will have a second chance. We all make mistakes when we're young; we all step out of line. Those with fortunate childhoods and youth, like mine, usually don't face serious consequences: if we slip up, a safety net awaits us. Yet some aren't that lucky. Victims and perpetrators at once, but above all, young human beings with their entire lives ahead of them.

My birthday cake is a sweet and sticky lemon tart that pops out of one of the standard greenish-grey food packets. There are no candles to blow out, because my thirty-eighth birthday is not a good enough reason to strike a match in the open cabin. To make up for it, Terry decorated the tart with some coloured letters that say 'Happy Birthday', and he's obviously rummaged through the bag of decorations to find a gaudy banner and a little hat for each of us.

We celebrate in the Service Module, now home to Gennadi and Misha, who arrived with Scott about a month ago. Gennadi, as usual, has our small party laughing boisterously at anecdotes from his previous spaceflights. I can't be sure, but I suspect that some of them are actually true, or that they contain at least more than 10 per cent truth, the minimum imposed on combat pilots in entertainers' clothes by a universal, if unwritten rule. Gennadi has a wide repertoire to draw from, since this is his fifth mission. A skilled and well-respected cosmonaut, next June he will beat the record for the longest total stay in space. When he goes back home in September, he will have made history by having accumulated more than 800 days in orbit. His outrageously irreverent humour won't make the history books, but his odd habit of never cutting his hair in space may get a footnote.

Alongside Gennadi and his extravagant personality, Misha can sometimes seem timid, with his sensible, contained nature and his kind and sedate ways. He's completely lacking that touch of callousness most of us astronauts have developed, at least outwardly, though it may not necessarily correspond to a genuine disposition. It was Misha who brought a recording of the sounds of nature on board, so that now, on weekends, as we float through the Space Station and pass within range of various computers' speakers, we can imagine we're in a rainy meadow, or, a bit further, on an ocean shore, and then in a forest ringing with birdsong. Before becoming an astronaut, Misha was a policeman, a job Scott knows well, since both of his parents were police officers. The two men will have a chance to explore this and many other subjects, since both will be on board for an entire year, a shared destiny they were anxious to celebrate when they arrived a month ago, and squeezed through the Soyuz hatch together to join us in the Space Station.

A surprising birthday gift from Earth arrived with them. I'm not sure how he did it, since it seems like an impossible undertaking, but Lionel somehow managed to get some packets of food made especially for me included in the Soyuz cargo. While I was training in the States, he went on a top-secret mission to Italy and along with Chef Stefano, now friends with both of us, he developed some dishes just for me. We ate together, he on Earth and I in space, during our usual video call. The menu? Delicious monkfish in a delicate, creamy sauce, broccoli and almonds, and red rice with asparagus tips. This is without a doubt the most sophisticated meal ever consumed in space!

This evening, we celebrate with the usual packets and cans. And thanks to the recent arrival of Dragon, the party is enriched by a rare pleasure: ice cream. Not only do we have a little stockpile on board, but today we can gorge without guilt: Mission Control actually asked us to eat all the remaining ice cream before tomorrow, since they need to make room in the freezer for scientific samples. As trained and fearless astronauts, aware of the gravity of the situation but not the least bit daunted, we assured them that the mission would be accomplished, cost what it might. Science is worth some sacrifices. In any case, we'll be able to compensate for the evening's excesses over the coming days. We're expecting the arrival of a Progress cargo vehicle, which will bring us the usual modest but welcome delivery of fresh fruit and vegetables.

The party ends reasonably early, because tomorrow is a workday, and we don't want to stay in the Service Module too late, since Anton and Gennadi sleep here. Terry and I are night owls, so we go to the Cupola as usual. It's dark, and we keep the lights off, helping each other find the right lenses from among the many that are stuck with Velcro under the windows. Scott has retreated to his crew quarters. Terry and I have a sense that our time in space is quickly slipping away, while Scott has long months ahead of him. Since his arrival, he's adopted the healthy approach of the marathon runner, someone who knows how to preserve his strength for the long run. He started his mission at a measured pace and, having been on the ISS once before for six months about four years ago, he was immediately ready to get to work both efficiently and autonomously. He also gave us some useful suggestions. As an example: Terry and I never questioned the instruction of closing the access to the Lab window with a rigid panel. Thanks to Scott, we were authorized to cover it simply with a curtain. Now, much more often, we can see a pair of legs sprouting from the floor, Scott's in particular, since he loves to take pictures with long telephoto lenses, and it's impossible to use them in the Cupola because of the distorting effects of the window scratch panes.

Another new thing Scott has taught us is to pay particular attention to the concentration of CO_2. The carbon dioxide we produce constantly by breathing is removed with special equipment, but the air in the Space Station is definitely not what you'd breathe in a forest: CO_2 levels above 3 millimetres of mercury are the norm up here, and that's a hundred times more than the average on the Earth's surface. Yet until Scott arrived, we'd never really paid any attention to it. The first few weeks on board, I had a few suspicious headaches and persistent tiredness, especially at the end of the day, though nothing that would get in the way of my working efficiently. Above all, nothing that could be definitively associated with elevated levels of carbon dioxide. Scott, on the other hand, is practically a human CO_2 sensor, and he associates specific symptoms with different levels above 2 millimetres of mercury. So Mission Control is doing everything in its power to reduce the concentration of CO_2, activating two scrubbing units simultaneously whenever possible. One day, we went from almost 4 millimetres of mercury to little more than 2 within a few hours. Despite the fact that I'd never had particular problems from higher levels, the breath of fresh air made me conscious of how stale the air usually is on board, a little like the way in which a sudden silence makes us aware of a background noise we'd got used to without noticing it before. I don't think that the excess carbon dioxide has affected my cognitive abilities, and to judge by the battery of computer-based tests I take on the computer every month there's no evidence to that effect. But I understand Scott's worries, all the more so considering that he's going to be on board a lot longer than the rest of us.

Terry has gone to bed. I'm staying up a few more minutes to watch the sunrise. I smile, remembering the most amusing present I got today. A group of my friends, more or less the same age and spread across half of Europe, put together a hilarious video in which they're singing the theme song to a cartoon from our childhood, inspired by *Around the World in Eighty Days*. Up here, all we need is a little more than eighty minutes. To the east, the familiar blue glow is starting to move across the horizon, a wash of colour spreading from the point where dawn will shortly arrive. I begin closing the shutters on the other side of the Cupola, the end-of-day ritual. Nothing gives me such an immediate and physical feeling of being in space as the simple act of turning the knob and hearing a direct connection with the thick shutters that close like petals over the windows. If I turn it more vigorously, they'll close more quickly. If I turn too vigorously, I'll hear them hitting the metal skeleton of the Cupola. It feels as if my own hands were outside.

Meanwhile, the sea of stars is slowly drying up, from east to west. The Sun lingers below the horizon before making its appearance, like a solo tenor waiting for the chorus to leave the stage before he walks onto it. Then, majestically, he appears on the scene. With dawn in my eyes, I give in to tiredness and common sense. I close the last shutters and retire to my crew quarters to float in my sleeping bag. Lying in bed to sleep? I have a faint memory of it, but I don't miss it.

38.

International Space Station, 28 April 2015

The entire crew is gathered around the table in the Service Module for supper, but there is no hint of the usual banter. The conversation is serious, and it revolves around a single subject. We're recalling precedents, discussing theories, making conjectures. And we're all asking the same question: what next?

This morning, everything went to plan, at least initially. When we flew over Baikonur at around 7 a.m., a few minutes before the launch, I was already in the Cupola, ready to scan the horizon for some sign of the rocket. I didn't see anything, but that didn't surprise or alarm me, since my attempts to track the launch in broad daylight had little chance of success. Similarly, at the morning DPC, there was no mention of any problem.

So we were waiting for the Progress 59P to arrive by early afternoon after the usual four orbits around the Earth, a prelude to what is always a fascinating encounter between two space vehicles in the vast emptiness of space. Gennadi and Anton had already set up the TORU in the Service Module, to be able to take manual control in case there was a problem with the automatic approach, when CAPCOM called us with an update: Moscow had announced a change from the fast six-hour rendezvous to the slower, classic rendezvous in two days. Still, nothing particularly alarming about that. There are so many minor malfunctions that can disrupt a fast approach without compromising the arrival at the Space Station. So we adjusted our expectations for the docking in two days' time.

But as time went on, we began to suspect that the Progress

59P would never arrive. In a rush of official announcements from Moscow combined with unconfirmed information obtained by our Russian colleagues, an account began to take shape. A catastrophic failure had occurred during the burn of the rocket's third stage. Gennadi had actually received a short video taken by the camera on the Progress. When he showed it to us a little while ago we could see the Earth quickly appearing and disappearing from the field of view, the sign of a rapid rotation. The theory is that one of the tanks might have exploded. With fading hopes, TsUP is persevering in its attempts to establish contact with the vehicle in order to obtain telemetry and send commands. But by this time it looks like the Progress has been lost.

How will this loss affect our mission? It's not as if they're going to make us come back early to save food supplies, is it? I've been on board for more than five months, and there are barely two weeks left before my departure, but the idea of having to give up even one day in space is distressing. Yet it's easy to persuade myself that it's silly to worry: a few days more or less will hardly make a difference to our supplies, and certainly wouldn't be enough to justify the enormous logistical effort involved in bringing our landing forward, with the associated deployment of staff and resources. Instead, I ask myself tentatively – not putting too much stock in it – doesn't this accident mean there's a chance that our re-entry will be postponed, as Anton and I have been joking? He gives a sly smile, saying little beyond, 'Who knows? There may be a chance now . . .'

International Space Station, 6 May 2015

Attempts to regain control of the Progress were abandoned after a couple of days. The cause of the accident is still unclear, but the chief suspect remains the Soyuz rocket, the very one that launches crews bound for the Space Station, though with a different third stage. It's therefore more than wishful thinking to suppose that the incident may have an impact on our return, since it seems inevitable that Oleg, Kjell and Kimiya's launch will have to be postponed to allow time for a full investigation; for the implementation of possible corrective measures; and possibly for the launch of another cargo vehicle before sending a crew to the launchpad. In this scenario, if we went back to Earth within a week as planned, Scott would be the only non-Russian crew member for some time, which would lead to an undesirable slowdown in activities. As it happens, shortly after the incident we had confirmation that NASA would like to postpone our departure.

However, it's not at all clear that it's possible. What is the situation with the provisions, now that we've lost a second supply vehicle barely six months after the Cygnus incident? What if the existing supplies of food and various frequent-use items were insufficient for a crew of six; would that rubber-stamp a 13 May return? From oxygen to wet wipes, from water to bin bags, from filters for the urine funnel to toilet inserts: Houston keeps track of these and many other consumables. In the hours immediately following the incident, the specialists promptly got to work to take stock of the situation. The verdict? Believe it or not, what seems to be in shortest supply are the solid waste containers from the toilet. When consulted, we didn't hesitate to say that we'd cut back and eke out the current stock. When it comes down to it, we'll just have to squash everything down, and we still have lots of single-use gloves.

However, the shortages on the Russian side seem to be more serious, and the advantages of extending our mission much less obvious, since in any case two cosmonauts would remain on board. So the question of our re-entry remains unresolved. It seems every day starts out with the promise of a decision, but the meeting of the Russian committee, which should make an announcement on the cause of the Progress accident and accept or reject NASA's request for an extension, is regularly postponed. There's a sense of stagnation on the Space Station, of unfulfilled expectation.

Despite the lingering uncertainty, we naturally continue to

prepare for our re-entry, and all the required tasks are put on our timeline as if there were no doubt about a 13 May landing. For a start, we study. Six months have gone by since the exams session in Star City, when we demonstrated how well we had mastered a nominal re-entry, as well as any emergency. It's time to refresh our knowledge. The thick descent checklist with its green cover and the malfunctions one with a red cover have both kept me company in my crew quarters for several days. It's a little risky, since if there were an emergency evacuation I'd have to remember to come and get them, but I want to take advantage of these snippets of time to refamiliarize myself with procedures, so I can face the return to Earth confidently. During some of our study sessions we got to hear the friendly, reassuring voice of Dima once again. Our young Soyuz instructor was at TsUP and ready to answer our questions after all this time. Anton and I also tackled a couple of manual re-entry simulator sessions, using custom software to practise taking control of the descent module and manually guiding it through the atmosphere to the designated spot for parachute deployment. We were definitely a bit rusty, but still comfortably within acceptable margins.

Continuing our preparations for re-entry, we put on our Sokol suits today for the first time in many months, and with bittersweet emotion. We strapped ourselves into our Soyuz seats and did a leak check on the suits to confirm that, in case of an emergency depressurization, they will be ready to save our lives. On a more pedestrian level, we've all confirmed that we still fit in the seats, despite a few centimetres of extra height acquired here. We'll lose them soon after landing, when the return to gravity will recompress the discs in our spine. Putting on the Sokol, taking in the unmistakable odour of the Soyuz, struggling to fasten all the straps, connect the oxygen and ventilation hoses, clumsily flipping through the checklists with thick gloves, and becoming one again with our little spaceship took me back to the whirlwind of emotions surrounding the launch, the overload of sensory stimuli on that incomparable night. But it's all different now.

There's no trembling expectation of a voyage I've yearned for. Instead, I'm reluctant, and resigned to the inevitable end of a great adventure. Maybe because of all the uncertainty and the ongoing rumours of delay, I really don't feel emotionally ready for my extraterrestrial experience to end. There hasn't been any gradual leave-taking from space, no mental preparation for the return or anticipation of all the nice things waiting for me on Earth. I've packed my personal things only half-heartedly. More or less knowingly, I'm betting on our staying, and if it doesn't happen, I'll be bitterly disappointed.

International Space Station, 10 May 2015

Finally, the uncertainty has come to an end. A couple of evenings ago, we gathered around a radio panel and listened over a private channel to the long-anticipated announcement from our lead flight director. The decision has been taken, and our Russian colleagues have accepted the postponement of our landing. I looked at Anton with an irrepressible smile: they've gifted us with at least another month in space. As if seeking further confirmation of this confidential communication, I anxiously awaited the evening's DPC for updates on the next day's timeline. Hurrah! The pre-departure test on the Soyuz navigation system, scheduled for tomorrow morning, had been cancelled. There was no explicit reference to the postponement because we were using an open channel, and a public announcement of the delay had yet to be made by the Russians. But now I was sure I could in good conscience set aside all preparations for departure.

And in fact, there were things to do to prepare to stay. After supper, I went to get a couple of new t-shirts I'd used to wrap things in a bag ready to load on Dragon. Other than some residue from the duct tape, they were in perfect condition. A rapid inventory of my wardrobe convinced me that there'd be no problem as long as I was careful. I still have a small stock of bonus food, including my favourite dish, quinoa salad, and as if that weren't enough good news, I even found a tiny bottle of barely used olive oil, and some bars of dark chocolate.

What might have been a weekend of frenetic pre-departure preparations had become instead a calm, ordinary weekend of life in space, and I've invited everyone to come to Node 1 for a cup of coffee. Literally a cup. Not only have I installed the espresso machine and got it going, but I've fetched some special cups from Dragon, designed for weightlessness. They were the brainchild of Don Pettit, a multi-talented NASA astronaut of explosive creativity. Some years ago, while he was on the ISS, he improvised the first prototypes with semi-rigid plastic sheets and adhesive tape. Everyone knows that in space you have to drink from sealed packets with a straw. It's not only a question of containing fluids: when you bring a cup to your lips and tip it, it does not, in weightlessness, cause the liquid to flow towards your mouth. The liquid stays stubbornly attached to the walls of the container – unless it's tapped, and doing that can set some of the beverage flying, forcing you to intercept it in mid-air with an open mouth. The space cups, though, are drop-shaped, with an acute angle precisely calculated to encourage the beverage towards the edge by capillary action. We can drink our coffee, then, by mimicking the gesture of drinking it on Earth, and more than that: we can savour the aroma. Who knows? Maybe one day Don's invention will be considered the beginning of design for weightlessness.

International Space Station, 7 June 2015

We've made good use of our extra month on board. Thanks to great rescheduling work on the part of the ground team, these weeks have been extremely productive. Suffice it to say that we have moved PMM: we've removed a module on the Space Station from its position on the nadir port of Node 1 in order to berth it to the forward port of Node 3. In truth, the actual move was effected from Earth, through remote control of the robotic arm. But we had a lot of preparation to do. In effect, we had to move a room in our space home. On one day alone, I worked from morning to night to dismantle and rebuild an entire section of Node 3's intricate ventilation ducts to a different configuration, so that the PMM could receive fresh air in its new position. So many tools were needed that Houston advised me to take down an entire drawer of tools and carry it with me. Luckily, with six months' experience behind me, I am good at organizing my workplace, optimizing my movements and preventing things floating away from me. Back on Earth, I'll miss these days of manual labour. When I'm no longer an astronaut, it would be lovely to find work as an artisan.

During the work of reconfiguration, I found myself in the bowels of the Space Station more than once, behind a tilted rack and in direct contact with the curved hull, usually hidden, which separates us from space. It's very thin, but on the outside there's a micrometeorite shielding that protects us from the smallest bits of debris. When they hit the shield, they are pulverized and scattered. This way, when the debris reaches the actual hull, the impact energy is spread over a larger surface. Bigger micrometeorites, however, would not be stopped this way. The larger ones, say, more than 10 centimetres in diameter, can be observed by the ground surveillance network, and when there's a risk of collision, however minimal, the Space Station is moved to prevent the undesirable encounter. It hasn't escaped anyone's attention, though, that there's an intermediate dimension, too small to be observed and too large to be neutralized by micrometeorite shielding. That's why it's so important to train for the scenario of rapid depressurization.

My last Sunday on the Space Station is coming to an end. I've spent it recording short explanatory videos about life on board, and they'll be inserted into an immersive tour of the ISS, for which I've shot a sequence of panoramic photos at high resolution. I also found time to film myself reading some fables, and – for Towel Day – an extract from *The Hitchhiker's Guide to* *the Galaxy.** These will be my best-loved videos, along with my reading of a few verses from *The Divine Comedy* in the Cupola, and unfolding the flag of my friends of the WeFly! team, an aerobatic team formed by disabled pilots with an inspiring story of endurance and friendship. This extra month on the Space Station has truly been a great gift. I've also discovered that it's allowed me to beat the record for a continuous stay in space by a woman, a record of whose existence I had never really been aware, to be honest, and which I'm achieving now because of the dubious distinction of having been unable to return to Earth according to schedule – a circumstance that had made me happy, incidentally. Now though, the moment has actually come to leave.

I enjoy a few more minutes in the Cupola, more in tune with my slight melancholy than with the spectacle of the Sun plunging into the ocean. Suddenly, high above the horizon, I see delicate blue filigree. The noctilucent clouds at last. Now I know I'm ready to go home.

* Towel Day is celebrated every year on 25 May by fans of Douglas Adams' books, who carry around a towel as recommended by *The Hitchhiker's Guide to the Galaxy*.

39.

when every star shines out,
and the shepherd's heart rejoices—[*text missing*]
that's the way
the many Trojan fires looked, as they burned there
in front of Ilion.
    Homer, *The Iliad*

International Space Station, 9 June 2015

This time it's for real: we're going home the day after tomorrow. 'Home' is what we call the planet, which glows with brilliant colours outside my window, in the light of a very ordinary, middle-aged star on the edge of the Milky Way. I'm in the Cupola and I want to start saying my goodbyes to the Earth: to the regular geometries of the desert dunes; to snowy-white peaks; to the wind, which takes shape as clouds, and to the myriad green lights of fishing boats in southeast Asia; to the blinding sun glint on a polished sea, and to the lattice of condensation trails from planes criss-crossing the trafficked skies of Europe and North America; to the folds in the Earth, its fractures and craters. There are no rituals or verses for this, no gestures. This experience is too new in the history of humanity. Not a single poet has seen what I am seeing, not one has been moved by the spectacle I have in front of me.

As my departure grows ever nearer, how can I make sense of my feeling of imminent loss, not only of the Space Station and this extraterrestrial life, but also of the planet I'm getting ready to return to, the planet from which I've been separated, but which has become so familiar to me in its entirety during our tireless orbiting? My eyes are saturated with sublime beauty, steeped in the splendour of the stars, overflowing with moving sunrises and sunsets, hypnotized by the dance of the auroras; they've seen the Earth change its clothes with the alternating seasons, all the while accompanying it on its eternal course around the Sun.

No, not eternal. The Sun will die, and with it, the Earth. But if we consider the fleeting quality of one life and the brevity of human history overall, how can we distinguish the billions of years that still lie between us and the end of our solar system from eternity? And all the thousands of years that have passed between the legendary fires at the walls of Troy and the fires from burning oil wells which, seen from up here, pepper the profound darkness of so many deserts on moonless nights: are these really any more than the batting of an eyelid? The distance that separates us from the Earth can't be simply measured in kilometres – barely 400. It's measured in the difficulty of getting to space, even just to low Earth orbit, where we are. It's measured in details that vanish from sight. It's measured by the speed with which a place appears on the horizon, only to disappear after ten minutes. It's a distance that renders daily life on Earth ever more remote and downright frivolous, but which weaves a bond across space and time: a commonality with men and women living on the planet right now and an affinity with past and future generations.

Something unexpected abruptly brings me back to the present. The Sun has set, as usual hurling fiery orange arrows. The Space Station is wrapped in shadow, and Earth's cities are sparkling. But something is disrupting the quiet familiarity of this scene: just outside the Cupola, lights are shining. They're coming from our Soyuz, which is docked a few metres away, ready to take us back home in two days. How can that be? The Soyuz has only one white floodlight, invisible from here – not clusters of orange lights spread along its length. Those aren't lights, they're the small attitude control thrusters, and there's no reason whatsoever for them to be on. Of course, we activated them briefly yesterday, when Anton and I tested the navigation system before our departure, but one channel at a time, each for less than a second, the bare minimum required to confirm that they work. And the Space Station was placed in 'free drift', which means it wasn't actively controlling its own attitude, so it wasn't trying to counteract the thrusters during our quick check. Could I have missed the memo about a new test controlled from the ground? It seems impossible; it's no small detail. However, the thrusters have been on now for several seconds. I'm about to call Mission Control when an alarm sounds from the loudspeakers. I head for the nearest computer, seeking information, while Terry is already speaking to Houston via radio. It's clear that the Soyuz thrusters should not have ignited: the Space Station has lost attitude control.

Attitude control is one of the indispensable functions for any space vehicle, be it a satellite, an interplanetary probe or our magnificent, complex home in space. We have to maintain the correct attitude if we want to orient the solar panels towards the Sun, correctly execute an orbital correction manoeuvre or point the antennae to communicate with the Earth. So how serious is our situation? The view outside is reassuring: we're not spinning out of control. We will inexorably drift away from the programmed attitude, that much is certain, but it's clearly a very slow drift. In the short term we won't lose contact with the control centres, and this is already reason to feel relieved. It's so much better having specialists on the ground navigating through many pages of procedure to resolve the problem!

The other good news is that the Soyuz thrusters have shut down, and our little spaceship seems to have gone quietly back to sleep, as if feigning never to have caused such a muddle. On board, we hesitate to go back to our scheduled activities. Anton and I do a quick check to see how much fuel is left on the Soyuz, and we're pleased to see that only one kilogram was consumed. There's more than half a tonne left for the voyage home. In the helium tank, which is used to keep the fuel pressurized, there's only a small drop in pressure, which could be attributed to a normal fluctuation in temperature. But many questions remain. It's clear that the on-board computer set off the alarm signalling the loss of attitude control following gyroscope saturation, and after the gyros had been working assiduously to counteract the Soyuz thrusters, which they tell us were on for thirty-eight seconds. But what caused this accidental ignition? Was it just a mistaken command sent from Moscow during a routine check, as it seems, or could it be a software bug? If it's a software bug, could it cause problems during re-entry? Will we have to carry out additional training to prepare ourselves for specific malfunction scenarios? Will it actually be necessary to update the software? One question in this overall confusion soon stands out, and it's the one that concerns us most: will the specialists be able to shed light on the situation in the remaining two days before our departure, or will our landing be postponed yet again? We've spent almost 200 days on this mission, and the warranty period for our Soyuz is close to its expiration date.

International Space Station, 10 June 2015

Yesterday I spoke to my family. Some of them are already in Houston, some are still on their way, while still others have their tickets but haven't left yet. They will follow our landing tomorrow from Mission Control, and the day after they'll welcome me at Ellington airport, where NASA keeps its aircraft fleet. After it was established that the thruster firing was due to a wrong command sent from TsUP, and not to a problem with our Soyuz, our departure was confirmed. Our little spaceship is ready to take us home. This time I welcomed the news with relief.

Underneath, though, I'm still feeling melancholy about my imminent departure, that rather sombre feeling you get when you leave a place you've loved, and you know you can't go back whenever you want, possibly never again. Despite this gloomy backdrop, I'm looking ahead with curious enthusiasm. Not too far ahead, not to gravity, which is waiting for me on Earth, and not to the effort of supporting my weight. I'm looking forward, in fact, to the experience of throwing myself into the arms of the atmosphere.

I've heard so many of my colleagues talk about it, their voices and descriptions maybe occasionally somewhat over-dramatic. I'm finally going to feel for myself now what it's like to fall back onto the planet. Scott gave Terry and me his tantalizing take on the experience. He assured us that he enjoyed that phase of his first Soyuz mission so much that he would willingly offer himself for another mission on the Space Station, purely in order to relive the re-entry on Earth.

So that's how I want my space adventure to end. Having fun.

40.

The Melancholy of Departure
Giorgio De Chirico, oil on canvas

International Space Station, 11 June 2015

My alarm jolts me from sleep just after midnight. I've managed
to sleep a few hours, thanks to a small dose of sleeping meds, and
I'm ready to begin my two hundredth day in space. The last one.
Still half asleep, I put a swab under my tongue and start the four-
minute countdown. It seems like any other morning, I think as
the swab absorbs my saliva. While I'm coming to, I scroll through
the last few emails that arrived during this brief night: best wishes
for the re-entry, see you soon on Earth. Then, perplexed, I see an
email whose sender I don't recognize. How strange, especially on
my last day. I only receive email up here from people I've personally authorized. I read the mysterious message – and burst out
laughing. It's full of irreverent humour and signed by a popular
American television comedian. Scott has recordings of this comedian's satirical night show sent to him regularly, and since his
arrival in March, we've often watched them together.

Still laughing, I open the doors of my crew quarters. Scott,
who's also just getting up in the crew quarters next to mine,
knows immediately that his surprise has been successful. I am
so touched. In order to arrange this gift, he must have had to
ask more than one favour. We float together towards Columbus, both of us in a good mood, and I take a blood sample, one
of my last donations of bodily fluids to science. I'm also scheduled to collect further urine samples, and that will continue until
we close the hatch. Last night, in what must have been one of

the most grotesque Space-to-Ground conversations in history, Huntsville tried to get Scott and me to coordinate our toilet trips so that we went right after each other. This way we'd only have to open the freezer once to put in our samples. We looked at each other, amused, and promised to give it our best.

Around one, we gather together for the last time for the morning DPC. Then a quick breakfast, and we're back at work. I confess that I'd hoped the final hours before departure would be calmer. I'd like some space to process the conflicting feelings I'm having, to welcome and honour them. But the schedule is pretty full. After the DPC Scott and I are together again in Columbus, where the blood samples have completed their cycle in the centrifuge and must now be prepared for re-entry in the Soyuz. We're somewhat perplexed by the packing instructions for the samples, and we quickly realize that this will not be straightforward. The new Twin Study experiment, in which Scott is participating along with his twin brother Mark, requires a series of specific samples and the packing instructions are quite confusing. We're in for a long clarification call with Huntsville. It's not even two in the morning, and for the second time today, Scott earns my immense gratitude, offering to look into this on his own.

I go to collect the stem cell samples from the freezer, pack them up and place a green sticker on them. I take a photo, which will be downloaded in Houston and sent to Kazakhstan to facilitate identification during the retrieval of urgent cargo. I then float towards our little spaceship, where Anton takes the package. He's still very busy putting everything in the descent module, in the limited space under the seats and in nooks and crannies free of equipment – all according to strict instructions received from the ground, to ensure that the vehicle's centre of mass corresponds to the calculations for our re-entry. Smiling and calm as always, Anton assures me that he's got things under control and doesn't need any help. So I get my personal radiation dosimeter out of my trousers, I put it in the pocket of my Sokol suit and float away. I've decided to take advantage of Scott's gift of time to make a last tour of the Space Station.

It's a long, silent goodbye. I try my hardest to stabilize in front of me one of the mini video cameras we have on board, but I'm trying even harder to imprint every last detail on my memory, starting with all the simple things we use every day on the ISS: the omnipresent rolls of duct tape and Kapton tape attached by a strip on the handrail; the blue microphones fixed with Velcro next to every radio panel, metres of thin white cable floating sinuously from them; the green and cobalt blue transformer bricks for the laptops; video cameras and still cameras with all sorts of lenses. I start my tour in Columbus, with its large, square bags tied to the floor with bungees; because there are so many activities, and storage space is always scarce. Then it's JEM, the largest and neatest module on the Space Station, organized with a meticulousness that's entirely Japanese; and its small airlock, where I installed some mini-satellites with my own hands, and they're now out there, in open space. Node 2, with my crew quarters, now almost empty and sparkling clean: a few days ago I removed the lining and took down the ventilation system to hoover up a surprising amount of thick, grey dust that had accumulated in the ducts, fans, grilles and acoustic cladding. The Lab, which I love to pass through with one vigorous, well-judged push from both arms, arriving at the other side without brushing against a single thing – not the bicycle, nor the back-up robotic workstation, the CPR bench, the potable water dispenser. Node 1, with its sloping table that doesn't get in the way. The little Airlock, dark as always when it's not in use, with two large white suits inside, looming out of the darkness like sleeping, anthropomorphic machines, hooded so that their helmets won't get scratched. Node 3, with the T2 treadmill on the wall, and where my blue running shoes are no longer in their usual place, stuffed under a handrail. My beloved Cupola and the PMM in its new position, overflowing with white bags of every size that create nooks and crannies throughout the space; the torch in my pocket has often proved useful in there. And then on towards the Russian segment, with its small, round hatches and pastel colours – except at the spherical junctions where there's a whiff of heavy industry, rather than sterile space laboratory.

When I reach the end of my tour in the Service Module, I do a slow cartwheel. Just because I can. And soon I won't be able to. It's this that I want to imprint in my memory above all, in every fibre of my body, something that can't be filmed on video: the feeling of floating, the pleasure of moving without the least bit of effort, and of inhabiting space in three dimensions. The freedom to eat upside-down, on the ceiling. For no reason except that I can.

I transfer the video to a computer, knowing that it will be downloaded automatically to Houston in the next few hours. Without giving it much thought, I add 'Restricted' to the file name, a word used to identify films of a private nature which aren't for distribution. There's actually nothing private about a tour of the Space Station, but I feel like this was my moment, something personal between me and the magnificent space home that has hosted me for 200 days.

That done, I go around erasing all traces of my presence on board. From Node 3, I collect the remains of my personal hygiene items, and apart from things that could be useful to Scott, I throw the rest in the rubbish. I check to make sure the camera and iPad are fixed firmly to the wall of my crew quarters with bungees. They'll be Kimiya's when he comes on board. I throw away the sleeping bag and the two or three remaining items of clothing. All I have now are the clothes on my back.

At six, I meet Anton in the Soyuz to execute a few activation procedures and checks. We are gradually awakening our spaceship. Hatch closure is scheduled for seven. I take a few snacks and a couple of water packets in the orbital module. I put on my nappy, and somehow manage to close my trousers over it. I glance over Columbus one last time, as if to reassure myself that I've left it in order, though I realize that my time on board is over, and I can do nothing more. Over these 200 days I've felt responsible for the whole of the Space Station, but as a European astronaut I've certainly had special regard for this bit of Europe in space.

There's just time for me to go to my crew quarters to post one last farewell message on my social media profiles. I've prepared a photo showing me in the Cupola in a farewell gesture: 'So long, and thanks for all the fish.'* When I'm done, I log out of my PC. Over the next few days Houston will remove all of my data and personal settings.

At a little before seven, we meet in the narrow tunnel of the MIM-1 module.

We said our goodbyes last night during our final meal together. Now that we're under rigid time constraints and being observed by the video cameras, we exchange only brief sentences of thanks and best wishes. But our hugs are warm and tight. I know very well that I'll rarely see Gennadi, Misha and Scott in the future. Yet I'm sure we'll always be bound through affection and unquestioned camaraderie, something time and circumstance can't undermine, since we're all part of each other's memories, with our extraordinary shared experience in the background. More than the view of Earth from the Cupola, the pleasure of floating in the air, the satisfaction of carrying out everyday activities that accomplish the goals of many people's efforts and hopes – more than all that, I'm going to miss the privilege of being a member of the International Space Station's crew. My stomach clenches when I realize that though life on board will continue, I won't be a part of it.

I don't have time to linger on this thought. Anton, Terry and I are now in our Soyuz, and the hatch, already closed, is a clear reminder of a packed sequence of events that now demand our complete attention and will end in approximately six hours when we hit Kazakh soil. After the first leak checks, we help each other put on our Sokol suits. I get in and then I use my hands to keep still, and let Anton tie and knot. When the last hook is fastened on the suit, I wash down two final tablets of sodium chloride

* A quote from *The Hitchhiker's Guide to the Galaxy* by Douglas Adams and the title of the fourth volume in the eponymous series. In the novel, the dolphins are saying goodbye to the Earth before it's demolished.

with plenty of water, and that completes my dose: 5g in the fifteen hours before landing. This is the doctors' recommendation to ensure that my body retains water, partially remedying the drop in blood plasma volume. It happens to everyone in space and can provoke symptoms of orthostatic intolerance, the sense of vertigo that affects astronauts who have just returned to Earth when they try to stand up too quickly.

I fold my clothes any old way; they'll burn during the re-entry along with the whole of the orbital module. I grab a snack to eat while I wait for Terry and Anton to change, and then I push myself into my seat in the descent module. It occurs to me that these are my last moments of floating freely. Once I'm strapped in place, I won't move until we arrive on Earth.

Strapping yourself into the seats of the Soyuz is always an exercise in patience and contortion, even more so without the weight that keeps your body down. I pay particular attention to the straps that keep my knees in position and will stop them hitting the control panel at the moment of impact with the ground. Despite the physical effort of strapping myself in, I don't feel too warm yet, and I delay switching on the suit's fan, enjoying a few more minutes of relative silence. Even after 200 days of free-floating in the Space Station, I still feel comfortable in my seat, with its custom liner shaped to my body.

I pick up the checklist. The various sections are divided by coloured plastic tabs: 'Emergency Descent', 'Program 5', 'Ballistic Descent with Computer Failure'. I hope I won't need any of these pages today, that I'll need only those for the nominal re-entry, rather boringly called 'Descent Number 1'. Next to the critical events, I've pencilled in the times Moscow sent last night. The first is at 11.44.40, when TsUP will transmit the command for activating the navigation computer, our little spaceship's brain. I glance at the clock on the control panel display, noting that there are only twenty minutes left. The clock shows Moscow time. I haven't even left the Space Station, but I'm already in a different time zone. Soon I'll fall within the borders of a country, someone will bring me my passport, I'll be given a stamp in one airport and a stamp in another. I'll be a foreigner everywhere on Earth, except on one small continent which we fly over in ten minutes. How odd, when you come to think about it. I'll have to get used to it again.

Anton is still closing the hatch between the descent module and the orbital module when the navigation system starts up. I hear Dima's familiar voice from TsUP requesting an update. I confirm that the accelerometer and the angular rate sensors are on, that the propellant tanks are pressurized. All indications are nominal. As soon as Anton is strapped in, we do the leak check on the Sokol suits, and then on the hatch between the descent module and the orbital module. After the Soyuz separates into its three components, it will be this hatch that protects us from the vacuum of space.

The leak check takes a long time, a good twenty-five minutes. At a certain point, Terry says abruptly: 'Samantha, all I can say is this: it's shocking how strong gravity is!'

When we've finished all the checks and waited for what seems like a very long time, we hear the voice of the flight director from Moscow on the radio, as he takes Dima's place at the microphone.

'Astrey, at the scheduled time of 13.18.30, send the command to open the hooks.'

I've prepared the command already. All I have to do is press the execute button. When the time comes, I make the short countdown out loud: 'Three, two, one. Command sent!'

With a touch of the baton, I put an end to our mission on the International Space Station.

41.

What is beautiful reminds us of nature as such – of what lies beyond the human and the made – and thereby stimulates and deepens our sense of the sheer spread and fullness of reality, inanimate as well as pulsing, that surrounds us all.
    Susan Sontag, *An Argument about Beauty*

Soyuz TMA-15M, 11 June 2015

As soon as the hooks release their grip, the springs on the docking system give us a light push and a separation speed of around 12 centimetres per second. The undocking is barely perceptible. In the black and white image I see the hatch slowly becoming smaller. These first few minutes are critical. We're in the immediate vicinity of the Space Station and we need to execute the separation manoeuvres correctly to prevent any collision: two brief burns of the attitude control thrusters of eight and thirty seconds respectively. If the computer doesn't complete the burns as scheduled, we'll have to intervene promptly in manual mode.

The second burn pushes us along the ISS, under it. The last thing I see of the Space Station is the external Japanese platform – and then my home for nearly seven months completely disappears from view. We're now on a safe orbit. We'll continue to move away from the ISS passively, so we can turn off the navigation computer. In a little over an hour, we'll turn it back on to start up a new automatic sequence – the last one, which will send us plummeting towards Earth.

'I can see you on the Space Station's external cameras,' Dima says in a chatty tone from Moscow.

Anton jokes, 'I'm waving my wings, can you see?', as if he were flying an airplane.

Dima keeps it up. 'Yes, I can!'

Thanks to the radio relay offered by the ISS, we'll be in constant communication with TsUP until the inevitable loss of contact while we go through the plasma shortly before landing.

Scott, Gennadi and Misha call us to say goodbye. 'Fair winds and following seas,' is Scott's sailor's blessing. 'It's been a pleasure spending time up here with you.'

Gennadi jumps in: 'Samantha, you forgot your fleece!'

Oh dear, and I thought I'd erased all traces of my presence . . . 'It's for you, Gennadi. Just in case it gets cold!'

'Ah, thanks!'

We won't have much to do for a good hour apart from regular system checks, and then just waiting and joking around to pass the time.

'Are we there yet?' Terry mimics a whining child on a long car journey.

I offer to tell him a joke. 'So there's a Russian, an American and an Italian in a spaceship . . .'

Terry never stops reminding me to drink. I get him to pass me one of the water packets he keeps next to him, where there's a little space.

When we restart the navigation computer, the Soyuz immediately begins scanning the horizon with its infrared sensors to orientate itself, as always, along the local vertical. The waiting continues for us, punctuated only by Anton's radio updates with TsUP. When we are correctly orientated and ready for engine ignition, still some time away, Dima passes the microphone to Yuri Lonchakov, head of the cosmonaut training centre, for the traditional weather bulletin.

'Good morning, Anton, Samantha and Terry. How is the mood on board?'

'Good morning, Yuri Valentinovich. We're all set on board. How's the weather in our landing place?'

'There's a little wind, visibility greater than 10 kilometres, clear skies. It's warm, 28° C.'

With these conditions the search and rescue teams will have no difficulty in getting to us. The only danger, it seems, is posed by the lakes that have formed on the steppe over the past few days after a period of unusually intense rain. They've told us not to worry: the recovery teams are prepared for a water rescue. I suspect that this scenario is quite unlikely, but it's nice to know that the rescuers are prepared for all eventualities.

After what seems an interminable wait, it's finally time for the engine burn. Dima reminds us to provide detailed commentary.

'Astrey, three minutes to go before the braking burn. Anton, we're expecting a periodic report from you on the deceleration. Samantha, you'll report on the engine operating pressures.'

'TsUP, from Astrey 1, copy. We're back in the light. Through the periscope I have visual confirmation of the correct braking attitude along the local vertical. We're ready for the engine burn. Samantha, you have the control screen, I have the manoeuvres screen.'

It's a seemingly insignificant braking – not even 2 per cent of our velocity: we're slowing down by only 128 metres per second. That's it. This is all we need in order to touch down on Earth in less than an hour. The engine ignites on schedule, and with a dull roar. Months ago, during the rendezvous manoeuvres, I barely felt the little thrust, but after six months in space my body is sensitive to even such a slight deceleration. All my attention is focused on the operating pressures. I'm determined to spot any off-nominal value immediately. Anton keeps an eye on the deceleration and the orientation. When he states that we've slowed by 74 metres per second, we know that the die is cast: even if we turned off the engine at this point, there's no way we'd stay in orbit. By now we're on the route to re-entry. It's irrevocable, and if there were to be a failure at this stage, we'd absolutely have to complete the braking burn, resorting to manual mode or the attitude thrusters as necessary.

Four minutes have gone by. We are getting close to the end

of the burn and we prepare to intervene if the computer doesn't send the shutdown command as expected. In Soyuz jargon, this is called the 'main command', as it's so important to shut down the engine on time. If you decelerate more than necessary, there's no way to turn back. Fortunately, our Soyuz continues to work impeccably, and the pressures in the injectors fall exactly as planned, following the main command. I'm waiting now for a sequence of three alarms, and I'm ready to shut them off using my baton.

First alarm: activation of the thermocouples. In some failure scenarios, these thermocouples would start the separation of the Soyuz modules once a certain increase in temperature was recorded as we passed through the atmosphere. I turn it off. Second alarm: activation of the automatic separation sequence, which will conclude in twenty minutes with the explosion of the pyrotechnic charges. I turn it off. Third alarm: opening of the orbital module's pressure relief valve. Off.

'The pressure is going down,' I report to Anton. We let all the air out of the orbital module so that there won't be any thrusts at the moment of separation. The pressure in our descent module remains stable, an implicit reassurance that the hatch is not leaking. I see a lot of Earth outside the window, because we've lowered the 'nose' towards the planet: we're pointing down now, instead of flying streamlined along our orbit. It's a way of facilitating the distancing between components when our spaceship splits into three parts: however fond we might be of the orbital module and the service module, we don't want to see them up close once we've separated from them.

After the braking burn, our orbit lowers inexorably, and through our headphones we hear interference from Earth radio stations, a sort of serendipitous musical welcome. Through the window, the Earth looks closer and closer, and the proximity has an alienating effect. For months I've admired a distant bas-relief, in which every movement was masked by distance and blurred by speed. Now that I'm re-entering the scene, being close to it unravels the familiar canvas of the planet, breaking it up into details. After we've crossed the Atlantic, I catch sight of a familiar landscape: below us is the Namibian desert, striped with red and strangely near, but always unmistakable. It's the last bit of the planet I recognize from space. I know I won't have any way to look for other familiar places: we'll shortly enter the atmosphere, and events will follow one after another right until we hit the ground. I'll be with humans again, like an ant that rejoins her community on the anthill after a furtive flight on an eagle's wings.

'Three minutes until separation. Are you ready to go for a ride on the American mountains?' Anton asks cheerfully. Terry and I both know he means a rollercoaster ride.

'Anton, in Italy we actually call them Russian mountains.'

We lower the visors on our helmets. We've been wearing our gloves since leaving the ISS, and our suits are therefore completely sealed and connected to the oxygen feed: if the cabin should lose pressure, they'll automatically inflate. At the appointed time, the pyrotechnic charges explode with a series of dull bangs and in so doing separate the three components of our spaceship. The thermal blanket around the descent module also detaches. I've waited for that moment with my eyes fixed outside in order to see it fly off, any part of it, not wanting to miss anything about this extraordinary descent. I see it move off, hesitate and turn back towards us as if to go back to its place. Then it detaches again and disappears for ever.

We have no immediate confirmation of the separation, but we'll soon find out if it's gone well. Only the descent module has the shape, the distribution of mass and the heat shield necessary to withstand the extremely high temperatures of atmospheric re-entry. As we slow down, we'll transfer to the surrounding air the energy stored in our tremendous speed; the energy the rocket transmitted to us months ago; the energy from 300 tonnes of propellant, transformed into nine minutes of formidable thrust. Of course, the atmosphere has no clear borders. As we descend, it gets progressively less rarefied and offers more and more resistance to our course. When the deceleration crosses a certain threshold, our Soyuz determines that atmospheric contact has occurred and it begins the automatic guided re-entry. Our two command and control displays switch to the descent screen. It shows a delay of five seconds: the computer will have to compensate by flying on a slightly steeper trajectory.

I already feel forcefully pressed into my seat and I'm amazed to read a bare 0.8 Gs. Less than 1 G, therefore less than my normal weight on Earth. We'll get to about 4 Gs, and not just once. I remind myself that my body has not become more fragile; it's the same body that took 8 Gs in the centrifuge with no problem, even if my brain now registers a single G as an oppressive pressure on my limbs. As the deceleration slowly presses me into the seat, I tighten straps as much as I can before it reaches the point where it's difficult to move my arm. The light plays across our cramped cabin as the computer rolls us about our axis, as it must do to keep us on the right trajectory for our re-entry.

Our heat shield is cutting a swathe through the air, and the atmosphere closes again over us after our passage, like the sea over a sinking ship. The air is resisting our intrusion, slowing our course with the violence of a hundred hands pressing on my chest. Then it heats up until it flares and sparkles. Bursts of yellow, orange and vermilion flicker around the heat shield, wrap around our bell-shaped husk and dart past the window like flaming tongues whipped by the wind. The heat blackens the window from the outside, gradually muting the iridescent palette until only the faint glow of burning charcoal peeps through. We're a lump of humanity enclosed in a fiery arrow.

Anton continues to provide commentary on the course of our re-entry aloud, even though no one can hear him now. After about six minutes and a second peak of more than 4 Gs, the deceleration tapers off, and my breathing becomes easier. By now we've lost most of our speed, and before long the parachute will open, the last critical event in this crazy descent. The sound of the wind is louder and stronger, a whoosh of air that lashes what's left of our little spaceship. I've heard many times that the opening of the parachute is the most violent moment of the re-entry, even more so than the impact with the ground. To keep my head from banging or bouncing, I push hard to settle my neck into the curve of the seat and wait. For a few seconds there's an inexplicable popping, like hail on a metal canopy, and then the drogue parachute opens. After the first jolt, we're suddenly battered here and there for ten seconds before another, more violent backlash accompanies the opening of the main parachute. Further brusque, haphazard oscillations follow until the capsule stabilizes, and we continue our descent, hanging peacefully under a canopy of 1,000 square metres. We're at less than 10 kilometres in altitude, where passenger planes fly: we've gone back to being Earth creatures.

Anton reminds us to tighten our straps, and I keep pulling stubbornly, in case there might be a few millimetres to gain. Then I place the checklist against my chest and cross my arms over it, trying to keep my elbows from sticking out. Nothing should get in the way of the seats' movement when the shock absorber is extended at the end of a final, rapid automatic sequence. First, a valve opens with a hiss, putting our cabin in communication with the outside, and then what remains of the heat shield is ejected to expose the retrorockets that will soften our impact with the ground. The outside windows, now blackened, detach, allowing the light of this beautiful Earth day to storm through. And finally, the seats' shock absorbers suddenly extend, and in a fraction of a second, I find myself a few centimetres from the control panel.

We soon hear the voices of the search and rescue team over the radio. They've seen us and when we're close to impact, they begin calling out our altitude: 'Four hundred metres.'

I force myself to look ahead, resisting the temptation to watch the ground approaching.

'Three hundred metres.'

For one last time I make sure that the whole of my spinal column fits snugly into the seat.

'Two hundred metres.'

Arms tight across my chest, teeth clenched.

'Prepare for landing!'

A sudden spark outside the window, like a flash of lightning, accompanies the ignition of the retrorockets. And then, impact with the ground, abrupt but not rough. The capsule wobbles, as if it's about to fall over, before it finds its balance standing upright. We wait a few moments to ensure that we're standing still. We are: it's a windless, late afternoon in Kazakhstan. I quickly take stock of my physical symptoms and find that the only painful spot is my nose: the impact caused me to bang my head against the visor of my helmet. Nothing serious. I'm sure it's not bleeding. No big deal, I think. I expected worse.

Terry, Anton and I raise our visors and join hands, still gloved, as we welcome each other back to Earth. My arms are heavy, my body's a boulder. My head is on the verge of spinning. And yet from outside comes the smell of grass.

42.

> But it could be that we, who are earthbound creatures and
> have begun to act as though we were dwellers of the uni-
> verse, will forever be unable to understand, that is, to think
> and speak about the things which nevertheless we are able
> to do.
>
> Hannah Arendt, *The Human Condition*

Namibia, November 2015

Yesterday a guide taught us to observe small desert animals:
discreet, ingenious, tenacious, obstinately attached to the least
chance of survival. Could I have imagined this abundance of life
when I was admiring the Namibian desert from the Space Station? Or the day I saw it from the window of the Soyuz – my last
extraterrestrial glimpse before returning to walk on the planet?

I didn't walk immediately, that's for sure. When they got us
out of the descent module, which still smelled of burning, they
carried us to large camp chairs. Brigitte gave me a quick check-up
and some anti-sickness medicine as a precaution. I looked at the
sky, so ostentatiously blue. The steppe, unusually green. The incredible number of people busying themselves around us. I was
there, and I wasn't there. Part of me was chatting with the people
who were speaking to me, but another part still imagined herself
on the Space Station, clinging desperately to memories of sensations, details and the slight, measured movements. Already I felt
unable to evoke them with any precision. The faint breeze over
the steppe was blowing them away.

In the nearby camp tent, Brigitte helped me to take off my
Sokol suit, and someone else immediately came to take it away.

It seemed a brutal separation, but I was in no condition to protest. The weight of matter was oppressive. There was a military helicopter outside, waiting to take me to the nearest airport, at Karaganda. So I took my first steps, supported on both sides, and my legs wobbled like toothpicks trying to sustain a boulder.

There was a cot ready for me in the helicopter. The mattress seemed so hard to me: too much pressure on my skin, which was used to tentative contact. As the sun set, I fell into a light sleep, though I often resurfaced, not to the point of complete wakefulness, but enough to incorporate my physical sensations into my dreams. In them, I found myself on board an aeroplane, and my body weight, exacerbated by my altered perception, was interpreted by my dreaming mind as a continuous manoeuvre at a high G load. Half asleep, I swore at an imaginary pilot: can't you at least fly straight and level for a while? I woke up smiling. In over six months in space I hadn't had a single noteworthy dream – or at least none I remembered. The return to gravity, though, immediately embedded itself in my dreamworld.

In Karaganda, while waiting around, I passed the first proof of fitness for terrestrial life, only just managing to avoid the classic mistake that betrays a long stay in space and a recent return to Earth. Brigitte had lent me her mobile so I could say hello to Lionel, without the rush of the brief call from the landing field. When I finished talking to him, I stretched out my arm towards Brigitte, firmly intending to let go of the phone and give it a gentle nudge in her direction. I stopped myself just in time.

After a brief welcoming ceremony and a few tests for balance and orthostatic tolerance, Terry and I said goodbye to Anton, who was leaving for Moscow. A few hours before, we were together in space; now we were separating like this, with a quick hug, and we'd see each other only a month later for the debriefs in Star City. Terry and I boarded a NASA plane equipped with two small beds for us to sleep on. In Scotland, the first stop on our return to Houston, we took a few steps around the tarmac. My balance was already much better and according to Brigitte I no longer had the wide-spaced legs of a baby just learning to walk.

But it was so tiring! After circling the plane a few times I got back on it to sleep, and instead of stretching out on the bed in a controlled way, I ended up falling backwards onto the mattress, as if some giant hand had pushed me. My brain would need another few days to relearn how to correctly estimate efforts and recruit the right muscular strength.

—

As the cool of the evening falls over the desert, Lionel and I stand outside, waiting for an imminent Space Station flyover. I saw it one summer evening in Houston a little after my return, and after that I didn't look for it any more, perhaps unconsciously afraid of reawakening my nostalgia, or maybe just distracted by the complications of life on Earth. In the first few days my workload was fairly light and still some way from the whirlwind of duties that would overwhelm me in the coming months, but tiredness was my constant companion. I slept a great deal, and my heartbeat was elevated even at rest, as if it were an effort simply to exist. My buttocks were sore when I sat, the soles of my feet ached as I walked and painful blisters formed on my tender skin. Wearing a regular bra made me feel breathless. A test on the vibration plate four days after my return indicated that my balance was still affected. Only after my second test a week later was I given permission to drive.

There's the Space Station, a shining dot in the dusk, silent sentinel and guardian of six human lives in the hostile vacuum of space. Soon they'll welcome another of the Shenanigans – Tim – and the United Kingdom is all abuzz. The great adventure is about to begin for him; the memories are beginning to fade for me. I try to remember how it felt in my body to float. I try to feel a push from the handrail in my muscles, or that light and precise touch with my feet that allowed me to do a somersault in mid-air in Node 1 and end up upside down right in front of the soup container. I try, but my life on the Space Station already has the hazy contours of a dream.

I can't say if more than 3,000 orbits around the Earth left some trace on my body. I feel as though my joints have lost some of their elasticity and I have slight muscular aches more often, but if they aren't the product of false memory or perception, should I put them down definitively to the months of weightlessness, or are they instead the natural result of the passage of time? I'm also uncertain about how this extraterrestrial experience has changed my way of thinking and feeling. I'm definitely calmer, but maybe this is simply the peaceful satisfaction you get when your dream is fulfilled, and your anxiety recedes. I believe I'm also a little less prone to allowing the need to feed my ego to get in the way of recognizing our shared humanity in the people I meet. But couldn't this just be a hint of the wisdom that routinely compensates someone who's nearly forty for the inevitable loss of physical strength? I do feel more profoundly convinced of the need for us all to live on this planet not as pretentious, quarrelsome passengers but as members of a crew on the same spaceship, loyal and ready to roll up our sleeves. Yet maybe this, too, is only the consequence of living in my time, a time in which it's not necessary to go into space to understand that all seven billion of us humans – and soon many more – are interdependent.

I wasn't expecting any blinding flashes of intuition after my mission in space, and I haven't had any. I don't know any more now than I did before about the meaning of human existence or the presence of extraterrestrial life. All the same, I can't help but think, or at least sense, that this experience has brought about in me a certain heightened sensitivity, and prepared my spirit to understand what others may one day put into words. Virginia Woolf wrote that it's the fire of genius that renders visible those premonitions Nature has traced in invisible ink on the walls of the mind. I hope that women and men of such genius will soon voyage into space, so that when I read them, I can exclaim, 'But this is what I have always felt and known and desired!'*

Meanwhile, I'm waiting for my next mission, though without the anxiety that preceded the first one, and it's still a privilege for me to contribute to human exploration in space. Aside from valid utilitarian considerations, and the legitimate goal of making our species multi-planetary to ensure its survival in the event of an unlikely but

* Virginia Woolf, *A Room of One's Own*, Chapter 4.

not impossible catastrophic asteroid collision, space exploration is for me a great adventure of the human spirit. It is a shared experience that nourishes our noblest side, raising us above our pettiness and boredom.

When I have the opportunity to leave once again, I'll be ready to go, with new colleagues, maybe even a new spaceship. And who knows? Maybe one day, a new destination. In the clear desert sky, the Space Station disappears below the horizon. I look at the Moon, and allow myself to dream. Our human adventure has only just begun.

Mountains rest, outgloried by the stars –
but even there, time's transition glimmers.
Ah, nightly refuged in my wild heart,
roofless, the imperishable lingers.

    Rainer Maria Rilke, *From the Remains of Count C. W.*

Acknowledgements

But, you may say, did it really make sense, after sharing my experience in real time on blogs, social media, interviews and documentaries, to write a book? I hesitated, even before I knew how much anguished effort it would require of me. It's true, a book is a permanent record rather than a series of fragments, and it's whole, deep and rounded. But after I had recounted so much, and with an almost proselytizing fervour, I felt I was suffering from communication fatigue!

In the end, I owe my willingness to write the book to my Italian publisher Elisabetta Sgarbi, who contacted me even before my mission, assuring me that she would wait for as long as it took so that I could let the experience settle before I processed it. I don't think Elisabetta expected that it would take me three years, but she accepted every delay in delivering the manuscript without the least sign of disappointment, encouraging me instead to continue with undiminished commitment. Thank you.

My Italian editor, Chiara Spaziani, has accompanied me on this long adventure with intelligence and a light touch, asking questions rather than offering answers, making my priorities hers, indulging my love of detail. When the painful time came to make substantial cuts to the original manuscript, I'm not sure whether it was because of a lucky affinity between us or perhaps because of her amazing ability to remain in sympathy with her authors, but her suggestions were always easy to accept. Thank you.

My friend Michela Borzaga gave me practical suggestions with the sensitivity of someone whose everyday occupation is great literature, and who knows what an effort it is to write. Above all, she encouraged, consoled and criticized me in the noblest sense of the word, helping me to find the voice I was looking for. Thank you.

My illustrator, Jessica Lagatta, perfectly captured the spirit of this book, uniting precise lines and evocative colour with a dreamy look that is nevertheless anchored in geometry. Thank you.

Thank you to the friends who read early drafts of the book: Sara Rocci Denis, who carried the text around the world and read the whole with the attention and rigour of an engineer, unearthing much that was imprecise and unclear; Stefano Sandrelli and Manuela Aguzzi, who read an early draft of the first few chapters; their encouragement at that stage was important; Tommaso Ghidini, who enthusiastically offered to be my beta-reader, despite the burdens of a very busy job; Amedeo Balbi, Matteo Massironi and Francesco Sauro, who kindly agreed to do the fact-checking for those pages for which I sought the reassurance of subject specialists.

The English edition was a new challenge, but also an opportunity for further fact-checking and improvement. I would like to thank my colleagues at ESA and NASA who reviewed the English draft: Scott Segadi and Regan Cheney provided reassuring feedback; Alex Kanelakos, Jean François Clervoy and Simon Challis made many precious suggestions; Marsha Ivins reviewed the entire draft with her characteristic thoroughness, efficiency and attention to detail.

A particularly heartfelt thanks goes to Paolo Attivissimo: his painstaking, expert review really made a difference. Paolo is also the unnamed Trekkie friend who suggested the perfect occasion for the now famous photo impersonating Captain Janeway in space.

Finally, let me thank my Penguin editor, Casiana Ionita, who wanted this book to be published in English. I am grateful for her patience and her unconditional support and understanding on the long journey to a final English manuscript.

Thanks are also due to those who, although not involved in this work, still contributed to this account of the Futura mission, giving up much of their free time.

Among those not named in the book I'd like to mention Anne Grudzien and Paolo Amoroso, the latter also the compiler of the glossary. For two years, they painstakingly translated hundreds of blog pages. As time went on, they were joined by others: Carlos Lallana Borobio, Benjamin Weyh and Dmitri Meshkov.

Thanks also go to Gianpietro Ferrario, who coordinated the

work of the Italian ARISS volunteers during the Futura mission; and to Marco Zambianchi, Michael Sacchi and all my friends from the Italian Space and Aeronautics Association, for their indefatigable work on outreach, beginning with the amazing podcasts 'Return to Futura'.

My biggest thanks go to my family, and above all to my parents. I would never have gone into space without the opportunities they created for me at great sacrifice, or the freedom they gave me to set no limits on my ambition.

Huge gratitude goes to Lionel, my companion. This book has cost not less than a thousand hours of work: it has taken many evenings, many weekends, many holidays away from our family life, many hours in which a cup of tea suddenly materialized on my desk at just the right moment, many words unheard because my mind was running over a thought that was difficult to put into words. Thank you for having waited patiently for this project to come to fruition.

Finally, I'd like to thank our daughter, Kelsi Amel, who, though unaware due to her tender age, nevertheless helped me too, by being healthy, calm, cheerful and at ease with everyone, something I hope is an early sign of her trust in the goodness of humanity. But most of all, I'm grateful to her for having demonstrated right from the start a great talent, shared with her mother, in the underrated art of sleeping long and well.

Glossary*

AIRLOCK
On the ISS, a module or compartment that permits movement of astronauts or equipment between the Space Station and open space. When the internal hatch is closed, the airlock is isolated from the rest of the Station; the atmosphere is pumped out or vented and then the external hatch can be opened. There are four airlocks on the ISS: one for equipment alone, two for EVAs in Russian spacesuits, and one for EVAs in American spacesuits. (The latter is indicated in this book as Airlock with a capital letter.)

APFR (Articulating Portable Foot Restraint)
A portable device for securing an astronaut's feet during an EVA so that the hands are left free to work. It can be fixed to the robotic arm of the ISS or to designated interfaces along the Station. It has three independent joints that allow the user to find the optimal orientation.

ARED (Advanced Resistive Exercise Device)
A machine for resistance training on the ISS, which allows users to perform the most common weight-lifting exercises in weightlessness.

ATTITUDE
The orientation in space of a vehicle, astronaut or object.

ATV (Automated Transfer Vehicle)
An ISS cargo spacecraft with automatic docking capability, developed by ESA. ATV carried out five cargo missions between 2008 and 2015.

---

* *Compiled by Paolo Amoroso.* This glossary is intended to support the reading of the text and not as an exhaustive definition of terms. Where a term may have a wider usage, the definition here is restricted to its meaning in the context of this work.

BACK-UP
Reserve crew for a space mission. They receive the same training as the prime crew and substitute for them if one or all members of the prime crew are unable to carry out the mission.

BALLISTIC
Refers to a vehicle's non-guided, passive movement through the air under the effect of gravity and aerodynamic forces. The Soyuz spacecraft is capable of making a ballistic re-entry in an emergency.

BAR
A unit for measuring pressure. The atmospheric pressure at sea level is approximately 1 bar.

BDC (Baseline Data Collection)
Sample collection and baseline biomedical measurements taken by researchers on astronauts before and after a mission. The BDC allows researchers to evaluate the results of a scientific experiment by comparing in-flight data with data obtained pre- and post-flight.

BONUS FOOD
A small supply of food chosen by each ISS astronaut in line with their personal preferences and set regulations.

BOOSTERS
Usually refers to auxiliary engines that increase the thrust of a rocket during the first phase of its ascent.

BRT (Body Restraint Tether)
A semi-rigid tool used during an EVA to anchor the astronaut to the workplace and for attaching objects to the suit for transport. It's an adjustable arm consisting of a series of metal spheres on a guide. By varying the tension, the arm can be tightened or loosened. One end is fixed to the mini-workstation on the suit, and the other is free and can be attached to handrails or objects as needed for carrying.

CAPCOM (Capsule Communicator)
The name for the NASA Mission Control Center position in Houston responsible for speaking to the astronauts on Space-to-Ground radio.

CAVES (Cooperative Adventure for Valuing and Exercising Human Behaviour and Performance Skills)
ESA training course for astronauts conducted in a space-analogue cave environment over several days, training them to carry out exploration activities and allowing them to develop and exercise efficiency, cooperation and risk management as part of international crews.

CEVIS (Cycle Ergometer with Vibration Isolation and Stabilization)
Stationary bicycle for cardiovascular training on the ISS.

CHAIR FLYING
Training technique borrowed from the aeronautics world, where it is generally used by pilots preparing to execute in-flight procedures and operations. It consists of mentally executing a planned sequence of actions and checks.

CHUMPS
Class name for the group of American, Japanese and Canadian astronauts who took basic training together in Houston starting in 2009. The name suggests 'goofballs', and was jokingly given to them in a play on words when they suggested 'Chimps' instead.

CREW CARE PACKAGE
Psychological support package sent to astronauts on mission in consultation with their families. It can contain food, clothes, gifts, letters, photos, books, personal effects or other objects that contribute to maintaining close ties and overall emotional well-being.

CYGNUS
ISS cargo spacecraft developed by the American company Northrop Grumman Innovation Systems (originally Orbital ATK). It has been flying cargo on behalf of NASA since 2013 within the scope of a programme for supplying commercial transport services to the ISS.

DCM (Display and Control Module)
Control panel attached to the chest of the EMU suit which allows the astronaut to monitor the telemetry and send commands.

DEBOOST

A procedure for lowering the orbit of the ISS through engine ignition on the Station or on a docked vehicle. The reboost – a procedure for raising the orbit – is more commonly effected.

DESCENT MODULE

The part of the Soyuz spacecraft in which the astronauts travel during the launch, the re-entry and other critical phases of the flight. Bell-shaped and equipped with a heat shield, it is the only part that returns to Earth with the crew.

DOGFIGHT

Aerial combat, usually at close quarters.

DONNING STAND

An EMU suit support that allows astronauts, during their training in the pool, to put the suit on without supporting its weight. It is attached to a platform which is lifted by crane while the astronaut is in the suit and lowered in the pool. The two analogous structures in the ISS Airlock have the same name.

DPC (Daily Planning Conference)

Meeting held via radio between ISS astronauts and mission control centres. Each workday there are two DPCs, one at the beginning and one at the end of activities.

DRAGON

ISS cargo spacecraft developed by the American firm SpaceX. It has carried cargo on behalf of NASA since 2012 within the scope of a programme for supplying commercial transport services to the ISS.

EMU (Extravehicular Mobility Unit)

A NASA pressure suit for extravehicular activities from the American Airlock.

ENVELOPE

Portion of space in which an astronaut in a pressure suit can work efficiently with the hands (work envelope); or the portion of space in which it's possible for the robotic arm to activate its capture mechanism to capture a cargo vehicle (capture envelope).

ERA (European Robotic Arm)
A robotic arm for the ISS, developed by ESA and installed on the outside of the MLM module.

EUROCOM
Name for the position at the Columbus Control Centre, or Col-CC, responsible for speaking with the astronauts on Space-to-Ground radio, usually about activities in the ESA Columbus module.

EVA
Extra Vehicular Activity or spacewalk: activity executed outside a space vehicle by a crew member wearing a pressurized spacesuit. On the ISS the activity is usually concerned with installation or maintenance.

FLAME TRENCH
A trench or large hole located under the launchpad to guide the rocket exhaust away from the pad at lift-off.

FLUID SHIFT
A redistribution of bodily fluids commonly caused by weightlessness.

FPS (Fan/Pump/Separator)
Component of the EMU pressure suit which ensures the circulation of air and water for cooling and removes condensate.

FREE DRIFT
Flight mode in which a spacecraft does not actively control its own attitude and position, and therefore does not react to external forces changing those.

G
Here it refers to a unit of measurement for the perceived weight felt in the centrifuge or inside an accelerating spacecraft. 1 G corresponds to a person's or object's weight on the Earth's surface.

SHLEMOFON
Cloth headgear with integrated audio (earphones and microphone) for radio communication. It's worn by astronauts under the helmet of the Sokol suit.

G-LOC (G-force Induced Loss of Consciousness)
Temporary loss of consciousness caused by the flow of blood away from the brain and induced by intense acceleration in the head-to-foot direction typical of aerobatic manoeuvring or aerial combat.

ISLE (In-Suit Light Exercise)
Pre-breathe protocol adopted for use in the NASA Airlock on the ISS in preparation for an EVA.

IV (Intra-vehicular)
Describes the role of an astronaut who manages the operations preceding and following an EVA in the EMU suit, such as suit donning/ doffing, pre-breathe and depressurization / repressurization of the Airlock.

J-COM
Name for the position at the Control Centre of the Japanese Space Agency responsible for speaking with the astronauts on Space-to-Ground radio, usually about activities in the Japanese module JEM.

JETPACK
Propulsive system worn on the back and usually powered by cold gas thrusters; for example, the SAFER on the EMU suit for EVAs.

KURS
Navigation system for approach and automatic docking of the Soyuz and Progress spacecraft on the Russian segment of the ISS.

LCVG (Liquid Cooling and Ventilation Garment)
Garment worn by astronauts under the EMU suit for EVAs. Contains integrated ventilation tubes and around 80 metres of smaller tubes to circulate cooling water.

LEAK CHECK
A check carried out on a spacesuit or a module whose pressure is different from the surrounding pressure in order to verify that any air or gas leaks are within acceptable limits.

LOCAL VERTICAL
An imaginary line that joins a space vehicle in orbit with the centre of the Earth, assuming for simplicity's sake that the Earth is a perfect homogeneous sphere.

MCOP (Multilateral Crew Operations Panel)
Panel composed of representatives of the ISS partner Agencies and concerned with crew matters such as selection criteria, certification, training and mission assignment.

MICROGRAVITY
Condition of weightlessness induced by free fall. The term highlights the existence of small residual accelerations due to the finite extension of the concerned system and other disturbances.

MILLIMETRES OF MERCURY
Unit of pressure measurement commonly used on the Soyuz and the ISS. The atmospheric pressure at sea level is around 760 millimetres of mercury.

MINI-WORKSTATION
Tool support fixed to the EMU pressure suit for EVAs. It is a metallic structure secured to the front part of the suit at chest level.

MLM (Multifunctional Laboratory Module)
Russian laboratory module on the ISS which will be used to conduct scientific experiments, provide crew services and host the external European Robotic Arm (ERA).

NBL (Neutral Buoyancy Laboratory)
NASA training facility with a large pool in which astronauts wearing the EMU suit train for EVAs. Contains a real-size model of the non-Russian modules of the ISS.

NEEMO (NASA Extreme Environment Mission Operations)

NASA programme that sends groups of astronauts, engineers and scientists to an underwater laboratory for missions of up to three weeks. Experiments, technology tests and simulations of space operations are carried out there.

NITROX (Nitrogen Oxygen)

Mixture of nitrogen and oxygen, containing less nitrogen and more oxygen than normal air. It is used in scuba-diving as well as during EVA training underwater, to reduce the risk of decompression sickness.

NOMINAL

Refers to normal behaviour for spacecraft and equipment, or an event that goes according to plan, without breakdowns or emergencies.

ORBITAL MODULE

A habitable part of the Soyuz spacecraft which includes the docking system and the hatch that provides access to the ISS. It is jettisoned shortly before re-entry into the atmosphere.

ORLAN

Pressure suit for EVAs from the Russian airlocks SO-1 and MIM-2.

PARTIAL PRESSURE

Can be intuitively understood as a description of how much of a particular gas is present in a gas mixture. More precisely, it is the pressure that each of the gases in a mixture would exert if it occupied the entire volume of that gas mixture at the same temperature.

PAYCOM

Name for the position at the Payload Operations Centre in Huntsville responsible for speaking with the astronauts on Space-to-Ground radio, usually about NASA experiments.

PAYLOAD

Here, scientific research activities and technological development conducted on board the ISS.

PGT (Pistol Grip Tool)

Battery-operated screwdriver with a pistol grip used by astronauts in their extravehicular activities in the EMU.

PHASE ANGLE

During rendezvous, the phase angle is the angle between the position of an approaching Soyuz and that of the ISS measured in the Station's orbital plane and with the vertex at the Earth's centre. Assuming that the vehicles are on the same orbit, if the phase angle is close to 0°, the two are close; if it is 180°, then they are diametrically opposed. Intuitively, it is an indication of the distance between vehicles along the orbital arc that joins them.

PLASMA

A hot, ionized gas in which most of the electrons are separated from their respective nuclei.

POGO (Partial Gravity Simulator)

A mechanical and pneumatic suspension system used by NASA to train astronauts for EVAs. It partially balances the weight of the subject, allowing them to rotate and to move vertically and horizontally, thus simulating some aspects of weightlessness.

PRE-BREATHE

Process that reduces the amount of nitrogen present in body tissues. It's carried out in preparation for an EVA in order to diminish the risk of decompression sickness owing to the difference between the pressure in the spacecraft and the operating pressure of the suit.

PREP AND POST

Training simulation for operations that precede and follow an EVA in the EMU suit, such as suit donning/doffing, pre-breathe and depressurization/repressurization of the Airlock.

PRIME CREW

In the ISS rotation system, a crew that no longer serves as back-up for a launch preceding their own. They are usually in the final six months of training.

PSI

Unit of pressure measurement commonly used in the English-speaking world. The atmospheric pressure at sea level is around 15 psi.

REDUNDANCY

In engineering, the duplication of components or critical system functions to increase reliability through the presence of reserve elements. On a spacecraft, it can also be achieved by allowing the crew to take over manual control in case of automatic system malfunction, or by training more than one crew member to carry out the same duty.

RELATIVE VELOCITY

The speed of one vehicle relative to another.

RENDEZVOUS

A series of manoeuvres that bring a spacecraft into the vicinity of a second vehicle.

ROLL-OUT

Transportation of a rocket from the assembly building to the launchpad.

SAFER (Simplified Aid For EVA Rescue)

Emergency propulsion system affixed to the EMU suit. Allows an astronaut carrying out an EVA to return to the Space Station in the event of accidental loss of physical contact with the ISS structure.

SERVICE MODULE

Non-habitable part of the Soyuz, with the main engine and solar panels. It is different from the Service Module (in this book with capital letters), which is a module of the Russian segment of the ISS.

S/G (Space-to-Ground)

Radio communication channels used by astronauts to speak from the ISS to the Mission Control Centres.

SNOOPY CAP

Cloth headgear with integrated audio (earphones and microphone) for radio communication, worn by astronauts under the EMU suit helmet. So named because it resembles the aviator cap Snoopy wears in the *Peanuts* cartoons by Charles Schulz.

SOKOL

Pressure suit worn by astronauts flying in the Russian spacecraft Soyuz. An emergency suit, it is designed to sustain an astronaut's life in the event of atmospheric loss in the cabin and cannot be used in space outside a spacecraft.

SOYUZ
Russian spacecraft for crew missions in Earth orbit, with a maximum crew of three persons; or Russian rocket fuelled by kerosene and liquid oxygen which launches the Soyuz and Progress vehicles.

SSRMS (Space Station Remote Manipulator System)
Canadian-made robotic arm on the ISS, also called Canadarm 2. It is used for the construction and maintenance of the Station, to transport equipment along the structure, to support astronauts in EVAs and for the capture of the cargo vehicles HTV, Dragon and Cygnus.

T2
Treadmill on the ISS.

TRACK AND CAPTURE
Manoeuvre for approaching and coupling the robotic arm SSRMS to a spacecraft in free flight in order to capture it and attach it to the ISS.

TRIP TEMPLATE
Table summarizing pre-launch trips to the various international training centres for all crews assigned to a mission on the ISS. It is periodically updated by the schedulers in the partner space agencies.

TRUSS
Large lattice structure on the ISS to which non-pressurized components are fixed, such as solar panels, radiators and pumps for the cooling system and storage platforms for spare parts.

TsUP
Russian Mission Control Centre responsible for Soyuz and Progress operations and for the Russian segment of the ISS.

UMBILICAL
A bundle of cables and flexible tubes that connect a rocket to the infrastructure of the launchpad and ensure electrical, propellant and gas supplies, data transmission and audio communication. Also, a bundle of cables and

hoses that provide breathing gas, cooling water and radio communication to an EVA suit when training underwater.

UTC
Universal Time Coordinated: the international standard for indicating time; it corresponds to Greenwich Mean Time.

VACUUM CHAMBER
A room from which air has been extracted to create a vacuum or greatly reduced pressure. During ISS training, it's used to simulate realistic conditions for the activities in the airlock before and after a spacewalk and for testing the Sokol suit in pressurized conditions.

VOIP (Voice Over Internet Protocol)
A telephony system that uses the internet or similar web technology. Skype is one example.

WEIGH-OUT
The process of attaching weights and floaters to an Orlan or an EMU suit to achieve neutral buoyancy and neutral attitude.